PAPERMAKING AT HOME

The complete guide to the ancient craft of papermaking, a craft that combines the aesthetic pleasure of creating a practical product that is uniquely your own with the satisfaction of recycling and making use of waste materials.

PAPERMAKING AT HOME

How to Produce Your Own Stationery from Recycled Waste

by

Anthony Hopkinson

Illustrations by

Ivan Russell and Sylvia Hopkinson

THORSONS PUBLISHERS LIMITED
Wellingborough, Northamptonshire

First published 1978

ISBN 0 7225 0483 7 (paperback)
ISBN 0 7225 0489 6 (hardback)

Photoset by Specialised Offset Services Limited, Liverpool
and printed and bound in Great Britain
by Weatherby Woolnough, Wellingborough, Northamptonshire
on paper made from 100% re-cycled fibre
supplied by P.F. Bingham Ltd., Croydon, Surrey.

CONTENTS

ACKNOWLEDGEMENTS

So far in my unfinished venture into papermaking I have met many kind people who have helped me in one way or another. There are, in fact, too many to list but one or two at least I can name.

Jozef Manteleers is a development worker in Rwanda. He is a very busy man but he finds time to write long and informative letters about his experiments in small scale papermaking and particularly about his researches into local crops in Central Africa suitable for pulping.

John Mason is the author of *Paper Making as an Artistic Craft*, published in the conventional manner in 1959 but now sold by its author in his own binding and with samples of his own paper pasted in by hand, so that each copy is unique.

Robert Partridge is the founder and workforce of the Two Rivers Paper Company. He makes beautiful paper in the traditional way but using ingenious equipment he has designed himself.

J.H. Young is Senior Lecturer in Paper Technology at UMIST – the University of Manchester Institute of Science and Technology. Despite their responsibilities to the paper industry all over the world, he and his colleagues have always been ready to talk and advise about papermaking at a more basic level.

The Society for the Protection of Ancient Buildings were kind enough to lend a copy of their publication *Water Paper Mills* by A.H. Shorter (London, 1966) for the second illustration, a stylized picture of a water paper mill.

I must not end this short and incomplete list without thanking my family, and my wife especially, for allowing me to make paper at home and for allowing me to use the kitchen liquidizer, without which this book would almost certainly not have been written.

Note:
Where 'pint' is used as a liquid measure in this book it is an Imperial pint, equal to $1\frac{1}{4}$ American pints.

INTRODUCTION

Once when I was demonstrating papermaking, a spectator came up and said: 'Why bother! You can get it quite cheaply at W.H. Smith's.'

He had a point, of course, but he was obviously one of those unlucky people who get no satisfaction from handling hand-made things. I am sure he buys plastic sliced bread for his television snacks and I feel certain he wouldn't let any craft interfere with his nightly viewing. He is probably one of the 93 per cent in Britain whose principal recreation is watching television.

This book is meant for the others. You can watch television while you make paper but it doesn't make for good paper or for good television. In fact, one of the things I most enjoy about making paper is the fact that you have to concentrate and that helps you to forget your worries about the rent, or the mortgage, or your job or the state of the economy or whatever else it is that bothers you.

People often ask me how I became interested in hand papermaking. It all started a few years ago when I was in charge of a factory making self-adhesive labels. We used to buy in rolls of special paper consisting of a face paper coated with adhesive and backed by a layer of siliconized paper. The printing process produced a lot of waste; in setting up to cut the label shapes and print a design on the face paper, we inevitably had to run through quite a bit of paper before we had one good label. Altogether between 10 and 15 per cent of the paper we brought in was thrown away. The customers paid for this waste, of course: it was included in the price of their labels. What bothered me was that no one seemed to want the waste. We even had to pay the local authority to take it away.

This started me wondering about waste paper. I knew that some waste paper was recycled but why not more? And why not our waste label paper? I began to talk to experts and to correspond with them and learned a little about the business of recycling. I fairly soon realized why our self-adhesive paper waste was so unpopular: the adhesive on the face paper and the silicone treatment on the backing paper were fatal to the recycling process. Recyclers of paper are bedevilled by what they call 'contraries' and no contraries are more so than these essential ingredients of self-adhesive paper.

By this time, though, I was hooked on waste paper and I continued to correspond about it. I felt I ought to use recycled paper for my own letters, just to make sure people knew I really cared. My efforts to make my own recycled paper were not very successful until I happened to find an article on making paper by hand. It appeared in a trade journal devoted to the paper industry. I couldn't find the name of the author but I later discovered that the article was a version of *Papermaking as an Artistic Craft* which was written by John Mason and published in 1959. This book deals with paper made from rags and plants, not from waste paper, but I learned a lot from it.

There was still one more secret to uncover and it took me a while to do so. Mr Mason's otherwise excellent book does not mention the fact that the pulp from which a sheet of paper is formed has to be a solution of as much as 99 per cent water with only 1 per cent of the vegetable fibre which forms the paper. Once I knew this I was writing letters on my own home-made paper.

John Mason's book is well named. Making paper is a craft and it can be artistic. As with all crafts, there are days when nothing goes well, but for me, luckily, these have been few. Often, the paper which looks very unpromising when you first make it turns out quite well in the end.

What Paper Is

Paper consists of microscopic fibres of vegetable matter, usually nowadays from coniferous trees but occasionally from other plants such as straw, sisal or sugar cane. The tree or plant is treated chemically or mechanically to break it down into its basic fibres and remove the softer, fleshy parts of the plant. The fibres themselves then get further treatment, called 'beating', to roughen their edges so that they will hold together better when made into paper.

The fibres are mixed with water to form a pulp. This pulp is spread over a wire mesh which allows some of the water to drain away, leaving an even layer of pulp which is rolled and heated to remove the remaining water. Essentially, only water and fibres are needed for the process; no glue is required, though various materials are added in small proportions to improve the quality or the finish of the paper.

That is a very brief description of what happens when paper is made. The process is now usually carried out on a continuous basis by large and complex machinery but the principles are the same whether the paper is made by machine or by hand.

Advantages of Hand-made Paper

Although the paper we all use every day has been made on a

machine there is still a small demand for hand-made paper and there still exist a few mills to make it in the traditional way. Craftsmen, often second- or third-generation paper makers, turn out paper of the highest quality, far superior in many respects to that made mechanically. This paper is used for printing special editions of books and for drawings and water colours. Artists prefer hand-made paper because they know it will last for centuries as a result of the care that has gone into not only the making of it but also the selection of the raw materials.

There is another important reason why artists prefer hand-made paper: when paper is made on a machine the mesh which takes the pulp is in the form of an endless belt running at great speed. The fibres tend to arrange themselves to lie in the direction in which the mesh runs and the paper will be weaker in this direction. When paper is made by hand, sheet by sheet, the method ensures that the fibres lie equally in all directions. When the artist comes to dampen the paper it will remain free from distortion.

If you tear a sheet of newsprint you will notice that it splits one way quite easily and in a straight line and in the other direction with a ragged edge, which is an example of the effect of machine running direction.

With the weight of centuries of experience behind them, and working to the exacting standards expected by their customers, the few professional hand paper makers left are creating a product which no amateur could achieve. On the other hand, papermaking is a craft which will give lots of fun and satisfaction to those who take it up, even if their efforts fall short of perfection.

You will enjoy making writing or drawing paper which is uniquely your own and you will find that the craft makes no great demands on you. The equipment costs very little and takes up almost no space when not actually in use. You will not be working at something which is going to take years to finish; your paper will be ready to inspect within minutes of making and you can, if you are really impatient, be using it within an hour of starting, though I personally think it is better not to speed up the drying process to that extent.

1
THE HISTORY OF PAPERMAKING

Rags make Paper
Paper makes Money
Money makes Banks
Banks make Loans
Loans make Beggars
Beggars make Rags

Anonymous
(c. 18th Century).

Before the invention of paper, men wrote on a variety of different surfaces. These included stone, clay tablets and materials such as silk. The two most successful preparations, although they had considerable limitations, were papyrus and vellum, or parchment.

Papyrus is a reed growing on the banks of the Nile. As far back as 3500 B.C., the ancient Egyptians found they could use it to make a writing surface. They split the stems of the papyrus and laid the reeds side by side. Then they laid a second layer of reeds across the first in the other direction.

After further layers the slab of papyrus was dampened and hammered flat and finally polished until smooth with hard stones. The Greek and Roman empires made papyrus and it only ceased to be used when it was superseded by the arrival of the craft of papermaking in the Mediterranean.

The ancient civilizations of America and some Pacific islanders made writing surfaces similar to papyrus from the bark of certain trees.

The words vellum and parchment tend to be confused; they are also often used to name certain papers of high quality. To be precise, vellum is obtained from calf skin and parchment from sheep skin. To make them, the skins had to be stretched, scraped, cured and rubbed smooth with pumice stone. Eumenes, King of Pergamum in Asia Minor (197-159 B.C.) is believed to have been the inventor of the material and the word parchment is thought to be a corruption of the name Pergamum. If this legend is true, the development of parchment was a case of necessity being the mother of invention. The Pharoah of Egypt, keen to prevent the growth of the library of Pergamum, had forbidden the export of papyrus, then the only suitable writing surface.

Vellum continued to be used throughout the Middle Ages in Europe and some important documents are still printed on it. Special editions of books are often bound in vellum, though I have noticed that such books often have their covers bent out of the flat by the vellum surface.

The Discovery of Paper

True paper, like a lot of other great inventions, was discovered by the Chinese. The process of papermaking was patented in A.D. 105 by T'sai Lun, an official of the Imperial Court, and later generations worshipped him as a god in recognition of his achievement. In our own times, evidence has been found to suggest that coarse forms of paper were made in China a century and a half before T'sai Lun's patent but his place in history seems assured.

T'sai Lun took old fishing nets and ropes and beat them in running water. After a long beating he achieved an even pulp which he spread on a special screen. This screen was made from strips of fine bamboo, laid side by side and sewn together at intervals with thread. The screen acted like a sieve and allowed the water in the pulp to drain away but not the fibres won from the rope by the beating process. He pressed the pulp with a weight and left it out in the sun to dry. When all the water had gone, partly from pressing and partly through evaporation, he could peel the pulp off the bamboo screen in a sheet. Later he substituted plant fibres for rope and nets and boiled the fibres to get rid of the non-fibrous stuff in the plants. He also used silk thread for making paper.

Apart from the materials employed by T'sai Lun, the Chinese used rags for making paper; early this century, during excavations in the ruins of the Great Wall of China, fragments of rag paper were found which had been made within fifty years of T'sai Lun's patent. In the British Museum, there are examples of early Chinese paper which are as good as any made today and much better than early European paper.

The Start of Papermaking in the Western World

The Chinese had little contact with the rest of the world and they did not care to share their ideas with others. It was not until A.D. 755 that the secret of papermaking started to spread westwards, and even then it was largely a matter of chance. At the siege of Samarkand, the Arabs captured some Chinese papermakers and learned the art from them.

The Japanese and Koreans probably started making paper just before this. Japanese hand-made paper is still famous for its quality. Immense trouble was taken over the selection and preparation of the pulp, which usually came from the inner bark of the mulberry tree. The Japanese developed paper so strong and flexible that it was used for making clothing and screens which were used as interior walls in their houses. In some mountain villages papermaking was the main occupation during the winter, with plants being beaten and washed in the streams that ran down the main street.

It was natural that the Arabs, having found out how to make paper, should take the process to Europe. At the time they acquired the secret from the Chinese they dominated the Mediterranean, though it was another 400 years before they established the first paper mill on European soil. This was at Xativa in Spain and it was founded in 1150. Papermaking next spread slowly through Italy, France and Germany to the Low Countries. It finally reached England at the end of the fifteenth century.

Papermaking developed in India and other Asian countries on its way to the Mediterranean. A lot of paper is still made by hand in India, and Mahatma Gandhi in the 1930s encouraged village papermaking as part of his campaign to develop rural self-sufficiency. I have heard that in the back streets of Tehran in Persia waste paper used to be collected and pulped and made by hand into thick paper file covers for offices.

The authorities in Europe did not take kindly to paper at first. The monks, who produced almost all written works, were opposed to paper because the invention came from the Moslem Arabs; in addition, it was generally thought that paper would perish while vellum would last indefinitely. The Holy Roman Emperor Frederick II (1194-1250) wrote a letter which is the earliest extant German document on paper, but he also issued an edict ordering that all official documents should be written on parchment. Even in the fourteenth century Italian notaries taking their vows had to swear to use nothing but parchment.

Jost Amman. A papermaker in 1558.

The First English Papermill

In 1490 John Tate started the first English papermill near Stevenage in Hertfordshire. Almost nothing is known about this and other early mills in England. Most of them did not last long because they could not compete with imports from

Stylized representation of a water paper mill of c. 1662 with all operations crammed into one picture. *A* is the water wheel with its shaft (*B*) and cogs (*C*) driving the stampers (*D* and *E*). *G* is the vat and *F* the press, with the drying loft above.

From *Theatrum Machinarum Novum* 1662, G.A. Böckler

established mills on the Continent. A century later, Sir John Spielman started a mill at Dartford in Kent, still an important area for papermaking. Spielman was Queen Elizabeth's court jeweller and he was granted a monopoly which probably went a long way to ensuring his success. He is said to have had 600 men working in his mill, and he received his knighthood when King James I visited it. Unfortunately, no example of his paper still exists today.

The first American papermaker was William Rittenhouse, a German who had learnt the craft in Amsterdam. He emigrated to America in 1688 and established a mill at Germantown, Philadelphia.

Until the nineteenth century there were few changes in the method of making paper. Linen and later cotton rags were the raw material. The rags were allowed to rot in water for two or three weeks and then chopped up by women and children. It was a very unhealthy occupation because the rotten rags were very unpleasant and often harboured diseases. Papermakers were not popular with their neighbours and mills were often in remote areas, with employees living on the premises. After cutting, the rags were transferred to a stamping mill for pulping. The stamping mill had a row of shallow bowls, each containing a heap of rags rinsed by running water. A hammer pounded the rags in each bowl. An axle driven by a water wheel and bearing a number of cogs raised each hammer in turn and allowed it to fall on the rags in the bowl. 1 lb of rags needed 100 times as much water to cleanse them.

The 'Hollander' Beater

In the eighteenth century an unknown Dutchman invented a better method of pulping. Stamping mills needed fast-running water to drive them and there were no fast rivers in Holland. The 'Hollander' beater was developed to overcome this difficulty. It could be driven by a windmill or by ox-power. It consisted of a bowl of oval shape with a partition running down most of its length. On one side of the partition was a roller with metal bars fixed to its circumference. There were similar bars on a bed-plate lying under the roller. As the roller turned, the rags were torn between it and the bed-plate. This action circulated the pulp in the bowl; it passed through gaps at each end of the bowl and was given a regular treatment from the roller for a number of hours. At the end of the process the rags were completely broken down. The beater also roughened the individual fibres in the pulp; this helped the fibres to mat together as paper and determined to some extent the quality of the end product.

The Hollander could produce many times as much pulp as a stamping mill, though it handled the fibres somewhat more roughly. Purists would claim that hand beaten fibres made superior paper and there are machines in Japan designed to copy the hand beating process. For all that, the invention of

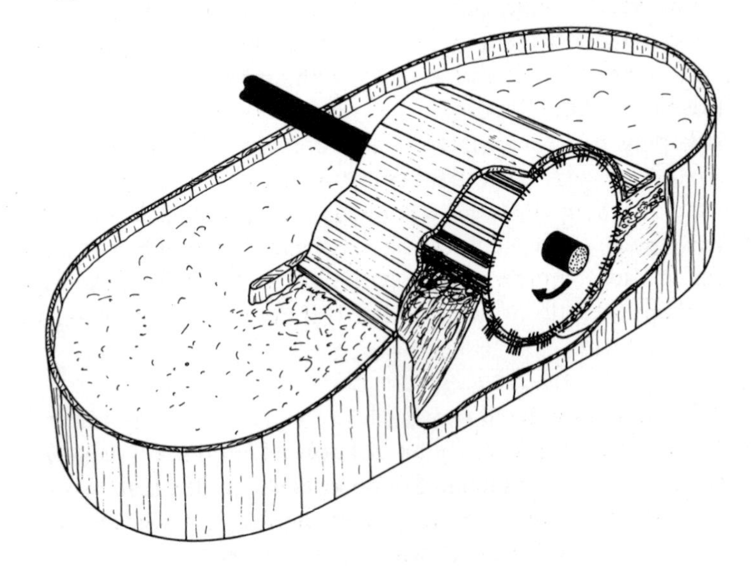

Hollander beater.

the Hollander was a very big advance and Hollanders are still in use today.

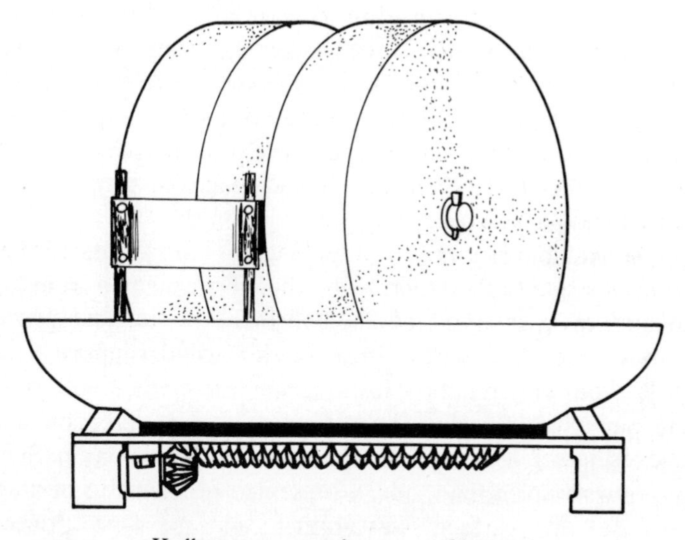

Kollergang – another type of beater.

Making the Paper

The pulp produced by the beating process was poured into a large vat, and the vatman, the most important worker in the mill, took his mould, a frame covered with a fine wire mesh, dipped it into the vat and scooped out a layer of pulp.

On top of the mould was a deckle, an open frame which kept the runny pulp on the surface of the mould until enough water had drained through the mesh to leave a firm, even layer of pulp, still 50 per cent water, on the mould.

The vatman passed the mould to the coucher (pronounced 'coocher') who turned the mould over and laid the pulp on to a sheet of felt. Another felt was spread over the pulp and a second layer of pulp couched on top of it. The vatman and the coucher continued making layers of pulp and couching them until they had built up a 'post' to a height of about 18 inches (45cm). The post was pressed to remove more water and the sheets of paper were separated from the felts and hung up to dry. Drying took place in a loft with lengths of cow-hair or horse-hair rope stretching from end to end. Sheets or batches of four or five sheets were folded over the ropes and the choice of cow- or horse-hair was important as other forms of rope stained the paper. Further pressing took place as the paper dried and it had finally to be inspected, sorted and packed ready for sale.

The mould was made by stretching fine wire across a frame from one side to the other, with other wires running from top to bottom at intervals of about 2 inches (5cm) and woven through the main wires. Strips of thin wood supported the mesh from underneath. This arrangement made a pattern in the paper of fine close lines running in one direction and thicker lines 2 inches (5cm) apart in the other direction. Such paper was called 'laid' and it presented problems to printers because the surface was rather uneven. One printer, Baskerville, suggested in 1750 the use of a woven wire mesh for moulds and this gave the paper a more even surface, called 'wove'.

The Introduction of Watermarks

Early in the history of European papermaking, the Italians were responsible for two important contributions. They discovered the use of animal glue (gelatine) for sizing the surface of the paper so that ink did not run. They also invented the watermark. This was a wire outline which was fastened to the surface of the mould. Where the lines of the watermark lay the paper was slightly thinner; this made a pattern on the paper which could be seen by holding the sheet up to the light.

Watermarks are a means of identifying the origin of a piece of paper. They were also a security device for bank note paper, making forgery more difficult. Despite that, bank note forging was a serious problem in Britain in the early nineteenth century. Sir William Congreve, a general in the Artillery and developer of the rocket as a military weapon, decided to defeat forgery with an elaborate variation of the watermark.

Congreve called his process 'triple paper'. A sheet of white paper was moulded and couched, and a coloured sheet couched straight on top of the white with no felt in between. A third sheet, white, covered the coloured sheet and when the three were pressed they formed a sandwich with white paper on the outsides and coloured inside. A stencil was placed over the mould when forming the coloured paper and this gave the coloured sheet a special shape. When Congreve's paper was held up to the light the shape of the coloured interior paper could be seen.

Congreve maintained that security was guaranteed because the coloured paper could be made from a special red cloth called Adrianople which came from a long and complicated dyeing process. Adrianople red cloth was only made at one or two factories and it would not be possible for the forgers to copy it. Alas, Congreve's ingenious idea did not find favour with the Bank of England and it was never taken up.

In the late nineteenth century the art of watermarking became very advanced. The original form of watermark, being raised, made the paper thinner and therefore lighter. If a depression was made in the mesh of the mould, the paper would be thicker and darker. A combination of depressions and raised areas would create a detailed picture with light and shaded portions. A portrait of the monarch could be created, or a picture of the paper mill.

In machine-made paper, watermarks are put into the paper in a rather different way. Instead of a pattern attached to the mould, the pattern is put into a mesh-covered roller running over the upper surface of the wet paper. The roller is called a 'dandy-roller' and its main function is to increase drainage during the early wet stage of papermaking. When a wire pattern is added to the dandy-roller the pattern will make an impression on the surface of the paper to give the watermark effect.

It has always been traditional to use blue paper for account books. There is a story, probably untrue, but a nice story all the same, about how blue paper was first developed. A papermaker was at his vat one day with his wife beside him doing her washing. Somehow her washing blue got into his vat and the first ledger paper was made.

Finding Raw Materials

At the beginning of the nineteenth century the raw material for paper was still rags and the colour of the paper depended on the colour of the rags. Only white rags could be used to make white paper. The discovery of chlorine and the development of bleaching powder made it possible to wash out colour and make white paper from any rags. However, there was a price to pay because the bleaching process introduced acidity which resulted in the paper deteriorating fairly quickly. It was not until much later that bleaching could be carried out without affecting the quality of the paper.

In the eighteenth and nineteenth centuries many experiments took place to find substitutes for rags as these were always in short supply. With the Industrial Revolution came the spread of literacy; more books were published and the development of newspapers also increased the demand for paper. It was vital to find another raw material. The first trials were of straw but by the mid-nineteenth century it was possible to treat softwood to make pulp and this is now the main source of all our paper.

Softwood is turned into pulp, either mechanically or chemically. In the mechanical process the wood is pressed against the face of a rotating grindstone and this breaks it down into fibres. The chemical process treats the wood to break it down into fibres and at the same time washes out non-fibrous material. Pulp made mechanically is inferior to chemical pulp because it still contains non-fibrous elements. On the other hand, the chemical process only recovers about 50 per cent of the tree whereas the mechanical process gives a paper weight equal to more than 90 per cent of the wood.

The Start of Machine-made Paper

Not only was it necessary to find new materials to meet the greater demand for paper. There had to be a new way of making it too because the hand process could not cope with the amount needed.

Papermakers were highly skilled craftsmen but tiresome people. They quarrelled amongst themselves and argued with the mill owners. Nicholas Robert, in charge of the workers at a French mill at the end of the eighteenth century, felt that the best way of ending the quarrelling was to invent a papermaking machine to replace the troublesome papermakers. His design was not successful and the French Revolution made further work difficult. It was taken to England and turned into a practical machine by an engineer named Donkin, working for two Fourdrinier brothers in the London stationery trade. Most modern machines work in much the same way as Donkin's, though of course they are now much larger and faster. They are still called by the name 'Fourdrinier' which is the only benefit the brothers got from their patronage; the cost of development nearly bankrupted them and the design was pirated so widely that they received few royalties from their patents. Brian Donkin continued to make paper machines for many years and several were exported.

In the Fourdrinier, pulp is poured on to a moving belt of wire mesh; as the pulp travels along on the wire, water drains through the mesh and more is removed by suction boxes placed under the wire. At the end of the wire section of the machine the pulp is picked up by a belt of felt on which it passes through rollers in the press section. The wire returns under the machine to the start and collects more pulp.

From the press section the length of paper, or web as it is called, passes through a bank of steam-heated rollers which dry and smooth the paper. A section of highly polished steel rollers called 'calenders' gives the final finish and the web is finally reeled up and removed for cutting into sheets or for other conversion processes. Much paper is delivered to printers still in reels for continuous printing. It is not cut into sheets or pages until the final stages of the printing process.

A modern Fourdrinier can turn out paper at a speed of 15½ miles (26km) per hour with a width of 8 yards (8m) but it still works in much the same way as Donkin's first machine, even if it now needs a computer to orchestrate its various functions.

In 1809 John Dickinson patented a paper machine working on slightly different principles. His invention has survived in the cylinder machine used today for making board. The most important part of Dickinson's machine was a vat filled with pulp. A wire-covered cylinder revolved three-quarters submerged in the pulp. There was an outlet for water connected to the interior of the cylinder. The water content of the pulp flowed through the wire face of the cylinder to the outlet, leaving a deposit of pulp on the face. At the upper part of the cylinder, clear of the vat, this deposit was picked off by a moving belt of felt.

Hand-made Paper Today

Machine papermaking advanced rapidly. By 1830, half the paper used in Britain was made by machine and by the end of the century hand-papermaking remained only for the production of special papers. Until quite recently it was the only way of achieving the strength and consistency needed for bank note paper and filter paper. Now there is only one commercial mill making paper by hand in Britain though here, as in the United States and on the Continent, there are individuals producing fine paper for special purposes in one-man mills.

In India and other eastern countries village mills making paper by hand are still quite common. I hope they will long continue, and I am personally involved with a move to introduce a form of small-scale papermaking, half way between hand making and machine manufacture, to many other Third World countries.

The one remaining commercial producer of hand-made paper is Barcham Green and Co Ltd of Hayle Mill, near Maidstone in Kent. The Green family have been papermakers since the end of the seventeenth century and John Green bought Hayle Mill in 1817. Now his brother's great-great-great-grandson runs it. One vatman at Barcham Green has been there for fifty years. Apart from the vatman and the coucher and layer who are concerned with the actual making of paper there are a beaterman, press operators and people to run the drying and glazing machines and sorting room.

It can take anything from two weeks to nine months, depending on the type of paper, to complete the whole process. Each sheet is handled about twenty-five times during manufacture and quite a proportion of the paper made is rejected for some defect at different stages of the operation.

Three main types of paper are made: *rough, NOT* and *hot-pressed. Rough* paper receives only one pressing; *NOT* (for some reason always written in capitals) is given a further pressing after removal of the couching felts to produce a smoother surface than *rough. Hot-pressed* paper does not actually receive heat treatment nowadays, but is placed between zinc sheets and passed through rollers. Contact with the zinc under pressure gives a very smooth finish.

One vatman making paper is backed up by as many as a dozen other people involved in the various handling processes. It will be obvious that we as amateurs cannot make paper to match the standard achieved at Barcham Green and its counterparts in other countries.

I mentioned one-man mills earlier. You will be very fortunate if you can visit one of these. A person making paper for others to use is totally devoted to quality and, being on his own, can take immense care at every stage of production to maintain this quality.

The Mill at Wookey Hole

There is one place in England where the craft of papermaking is permanently on view. At Wookey Hole Caves, near Wells in Somerset (once advertised as the oldest stately home in England), the River Axe emerges into a wooded valley.

The site, with ample pure water, was particularly suitable for papermaking. The first paper was made there before 1610. A family called Hodgkinson owned the mill and the caves for a hundred years from the mid-nineteenth to the mid-twentieth century. At the end of their time ownership of the mill and of the caves was split between two branches of the family. The mill was sold by its owners to the Inveresk Paper Company, which already owned another mill approximately half a mile downstream. Operations at Hodgkinson's mill were closed down by Inveresk in 1972. The following year the caves and the mill were reunited under the ownership of Madame Tussaud's.

Madame Tussaud's use part of the mill as a fascinating store for wax heads, plaster moulds and accessories of prominent people presently not required for display in the London exhibition.

It should be well worth a visit for any pop star or politician needing to be reminded of the brevity of fame.

Since Madame Tussaud's have revived the paper mill it is also worth a visit for anyone interested in papermaking. The various departments of the earlier business are on display and

John Sweetman, in charge of papermaking at Wookey Hole, has now built up an operation which is actually producing paper for sale. You can see the paper being made and you can buy the items produced there.

2

THE EQUIPMENT

List of Materials

Mould and deckle:

planed wood $1\frac{3}{8}$ x $\frac{3}{4}$ inch (35 x 18mm): $6\frac{1}{2}$ feet (2m) length

Mesh:

Nylon or bronze wire mesh, 30-50 holes per inch (12-20 per cm): a piece $15\frac{1}{2}$ x $12\frac{1}{4}$ inches (40 x 31cm). Nylon or terylene net curtain material can be used instead.

Metal corner plates:

You will need eight.

Screws:

$\frac{3}{4}$ inch (20cm): 30 (preferably brass)

Brass pins:

Not more than $\frac{3}{8}$ inch (10cm) long: 50

Couching cloths:

1 pack of mopping-up cloths

Press:

Blockboard: 2 pieces, size 14 x 10 inches (35 x 25cm) approx.

Hardwood battens: 4, $1\frac{3}{8}$ x $\frac{3}{4}$ inch (35 x 18mm), each about $11\frac{3}{4}$ inches (31cm) long.
Bolts: M10 x 110, that is, $\frac{3}{8}$ inch dia. x $4\frac{1}{2}$ inches long (10mm x 11cm), complete with wing nuts and washers: 4 G clamps can be used instead of nuts and bolts.

The Mould and Deckle

The mould is just a simple rectangular frame with mesh stretched across it. The deckle, an open frame of the same size as the mould, will lie on top of the mesh-covered side to turn the whole thing into a temporary sieve.

The moulds and deckles used by professional papermakers had mahogany frames made with special joints designed to stand up to years of daily immersion in water. They lasted for sixty years or more. Your mould and deckle do not have to withstand such treatment so you can make them from ordinary planed deal, though hardwood would last longer. I still have my first mould, made two and a half years ago from waste timber thrown out by a builder.

If you are an expert carpenter you can dovetail join the corners of the two frames; if not, you can make do with a simple butt joint. However, if you do use a butt joint it would be wise to reinforce each corner of the frames with a metal corner plate.

The inside dimensions of the frames will be decided by the size of paper you want to make. I suggest you start with a sheet of A4, which means $11\frac{3}{4}$ x $8\frac{1}{4}$ inches (300 x 210mm). That will then be the inside area of each frame.

When the two frames have been made, one of them has to be covered with mesh to form the mould. The best mesh to use is either nylon or bronze wire. Nylon mesh is used for filtration purposes and can be got from some silk screen printers' suppliers. Bronze wire is sometimes available from large hardware stores but it isn't easy to come by. A simple and easily obtainable alternative is net curtain material, usually either of nylon or terylene. Net curtain material unfortunately doesn't last very long and it is hard to get enough tension; it is likely to sag in the middle unless you support it with fine threads stretched across the frame under the mesh.

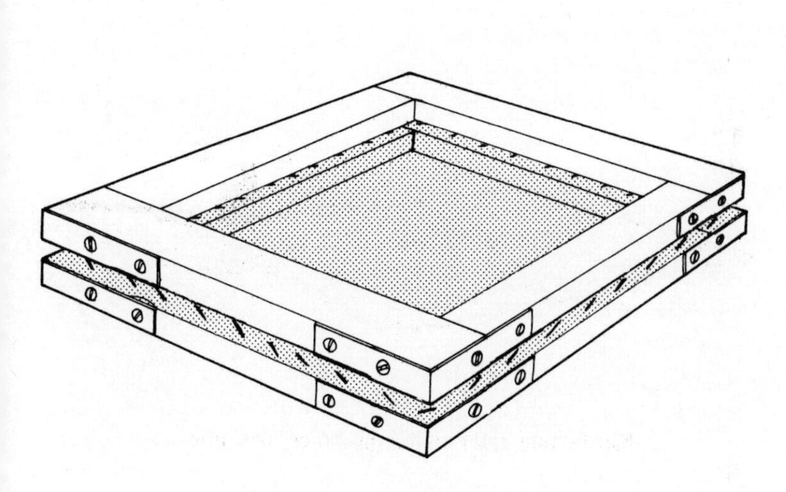

Home-made mould and deckle.

Whatever mesh you choose, it should have between 30 and 50 holes per inch (12-20 holes per cm). The professionals use a finer mesh than this but you will find it easier to use a coarser grade and the difference will not much affect your paper.

When you stretch the mesh across the face of the frame it must be as tight as possible. If you are using nylon, dampen it first because nylon stretches when wet. Use fine brass pins to fix the mesh because they won't rust. Work by putting a pin in on each side alternately, keeping the mesh stretched as you go. You will find this no trouble if you can use a pair of stretching pliers normally required for upholstery work.

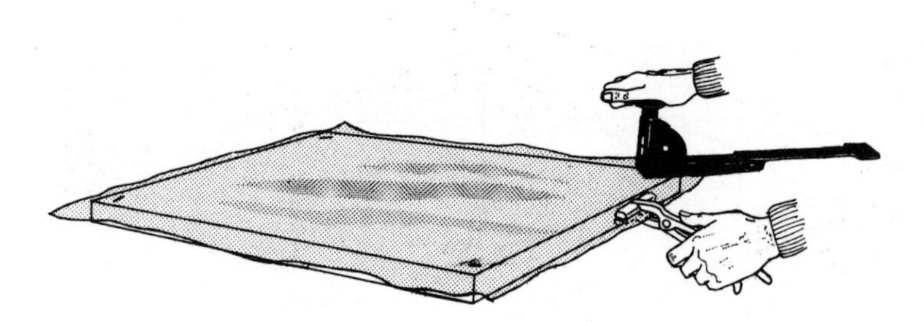

Stretching and tacking mesh across frame.

Another way to get a well-stretched mesh is to make a rough outer frame with inside dimensions just a little larger than the outside dimensions of your mould. Tack or staple the mesh to this outer frame. The object is to lay this temporary outer frame on top of the real frame and force the temporary one down so that the mesh stretches over the face of the real frame. If you put the mould frame on top of the deckle frame and lay both on a piece of board with a slightly larger area, you will be able to lay the temporary frame on top of the mould and force it down by clamping it at either end to the board on which everything is resting. Once the mesh is taut you can drive pins in to fasten it to the mould and cut it free from the outer frame.

Temporary frame for mesh stretching.

I have emphasized the need for a taut mesh but don't worry if your first effort has not been too successful. It will probably be quite adequate for your first making of paper.

The professional vatman has a deckle with a lip all around designed to fit snugly over the top of the mould and stop pulp from slipping out between the deckle and the mould. It also ensures that the paper has a neat edge to it. You can achieve rather the same effect later by glueing strips of felt to the face of the deckle so that it will make a tight fit with the mould.

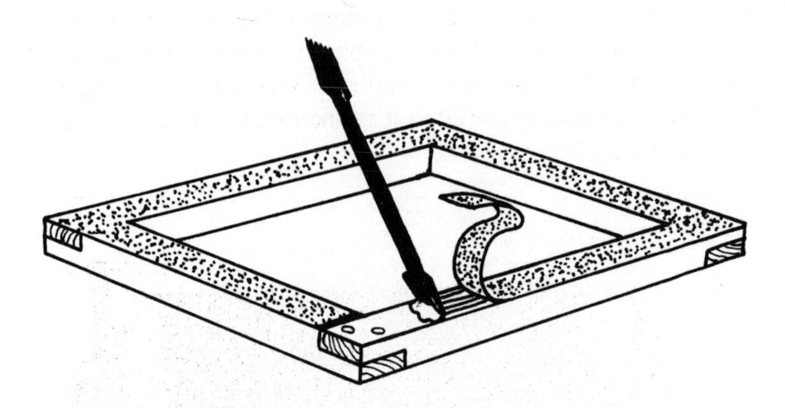

Glueing felt strips to face of deckle.

Couching Cloths

After each sheet has been made on the mould it has to be turned over and laid on a flat surface. This is the process traditionally known as couching; the word probably comes from the French *coucher*, to lie or lay down. A couching cloth is needed to cover each sheet before the next sheet is couched on top of it.

Couching felts were always made from woven wool but you can use any absorbent cloth. Old blankets are excellent, though they will not last terribly long. Perhaps the most convenient things to use for couching cloths are the synthetic mopping-up cloths sold for use in the kitchen. They are sold by grocers and supermarkets and one well-known make is called 'J-Cloth'. You should, however, avoid J-Cloths because they have holes in them which, though these make them all the better for use in the kitchen, spoil the appearance of home-made paper. The smooth ones without holes are excellent because they soak up water well and dry off quickly. They peel easily when you come to remove them from the paper you have made. Since they are not subjected to any wear in papermaking they last a very long time.

You will need about twenty-five couching cloths and you should cut them to size so that they are about 1 inch (25mm) larger on each side than the inside dimensions of your mould and deckle.

A Press

Modern mills making paper by hand have large hydraulic presses to expel as much water as possible from the post of couched paper. You will have to apply some form of pressure to your stack of paper for the same reason. The simplest form of press consists of a board on top and a board underneath, with the paper sandwiched in between. If the size of the sheets is small, say $8\frac{1}{2}$ x $5\frac{1}{2}$ inches (21 x 15cm), standing on the upper board will give enough pressure. For larger sheets more weight will be needed; perhaps two or three people standing on the upper board. If you are not part of a circus act you can put a plank across the upper board and line your family up on the plank.

There are, however, easier ways of pressing paper. The illustration shows a simple press made from two pieces of blockboard, four cross battens of hardwood and four nuts and bolts. This press works well and lasts a long time. You can substitute four G clamps for the nuts and bolts.

You may be lucky enough to find an office copying press for sale. In the early part of this century these presses were used in every office to make copies of important documents. Although the typewriter existed, letters of any significance were written by hand. When a copy was to be kept, the surface of the letter was dampened and the sheet laid face down in a bound book of blank pages. The book was put into the press and pressure applied to it. When the letter was peeled from the page of the book a perfect copy remained in the book. It was, of course, in mirror writing but there was no

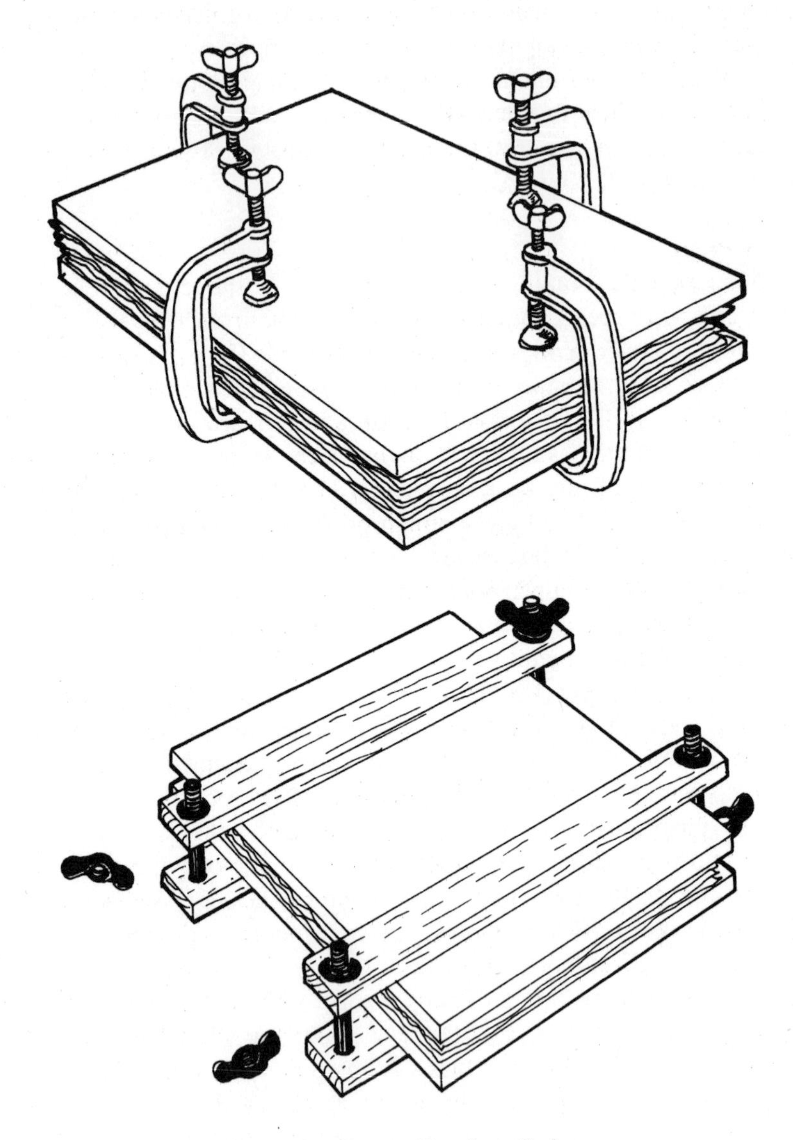

Two kinds of press. *Top:* four G clamps.
Bottom: bolted hardwood strips.

possible danger of losing copy letters. Office copying presses are not easy to find and, like so many relics of the past, they are attractive to collectors so you may have to pay quite a price if you do find one.

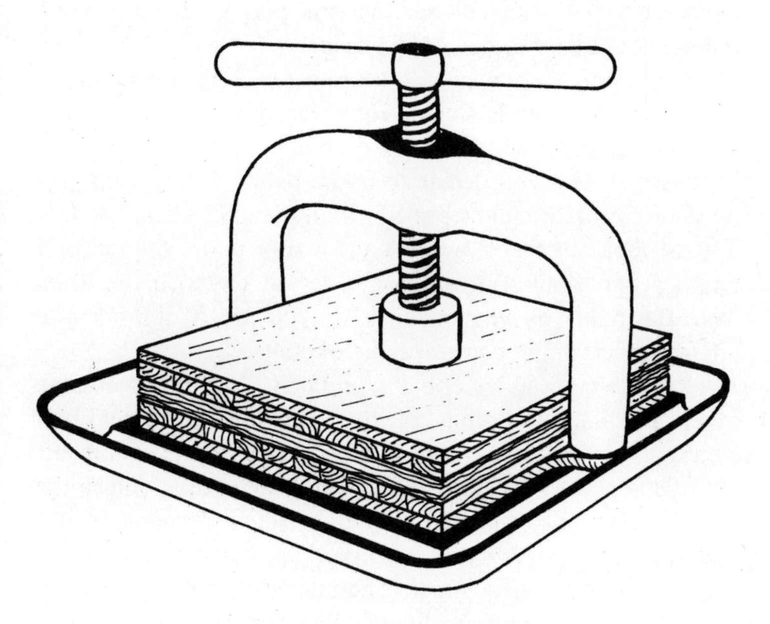

Office copying press, c. 1900.

The Vat

The vat is the container which holds the pulp from which you will be making paper. A kitchen sink will do very well as long as it is big enough to hold the mould and deckle. As an alternative to the kitchen sink any rectangular bowl of the same shape will do as long as it is not less than 5 inches (13cm) deep. Don't worry about the drains; if you use the proportions of pulp and water I will be recommending, the mixture will be very weak and far too runny to cause a blockage when you pull out the sink plug or pour the last remnants in your vat down the waste.

As so much water is needed for papermaking, it is bound to be a rather messy business. However, you will be spilling nothing but water so you won't do much harm to the kitchen or wherever else you decide to make paper. If any pulp gets on your clothes it can be peeled off quite easily when dry. It is a good idea, all the same, to have a tray under the vat and another flat, shallow tray beside the vat to catch the drips from the paper as you couch it. If you are working in the kitchen the draining board will do, of course.

That covers the essential equipment for all operations except pulping and drying, which are covered in the next two chapters. No special care needs to be taken of the equipment but it is important to clean the mould, deckle, couching cloths and vat after each making. Any pulp which remains behind and is allowed to dry will get mixed up with the next batch and will spoil the new paper. The couching cloths should be rinsed out in washing-up liquid from time to time to keep them fresh.

3
MAKING PULP

The easiest of all raw materials for papermaking at home is newsprint. Being unsized and short-fibred it breaks up more easily than other forms of waste paper. For these very reasons it does not make the best paper, so you will probably want to try various alternatives later, once you have taught yourself with the aid of this easy material.

Newsprint is printed closely with a heavy black ink; when it is pulped the ink will mix with the paper and turn it a greyish colour. In Britain, amateur papermakers are lucky enough to be able to use the *Financial Times* which is printed on pink paper. When it is pulped, the ink and the pink paper combine to make a brownish hue which is much more pleasant than the usual newsprint grey. I have heard that the quality of the paper used in the *Financial Times* is rather better than in some other newspapers. It is said to be less acid and paper which has a high acidity discolours with age and deteriorates relatively quickly.

To make the first batch of pulp, take some sheets of newspaper and soak them for an hour or more in water. You will need 5½ oz. (150g) which is about fifteen sheets of a full sized newspaper or thirty sheets of a tabloid. You can get rid of quite a bit of the ink in the paper by boiling it for about an hour in 6-8 pints (4-5 litres) of water with 2 tablespoonsful of detergent. Remove any scum which floats to the surface during the boiling as this is the ink which has been released from the paper.

Newsprint in Britain is non-toxic, so the paper and ink can be prepared in receptacles which may also be used for food without causing any harm. In other countries it would be wise to ask about toxicity. It certainly will not be very serious as

Boiling the pulp to get rid of the ink.

long as you take good care over washing the receptacles afterwards. Any black deposit left behind by the ink in the newsprint will be easily removed with detergent.

Pulping by Hand

Although pulping is much easier in a liquidizer (blender), you can do it by hand. Put the wet newspaper into a bucket and beat it with the base of a bottle or a thick piece of wood as if you were using a pestle and mortar. There should be enough water to make the pulp slushy but no more. The beating will gradually break the paper down into small particles and eventually into basic fibres. It is quite hard work and takes a fairly long time. You can test progress now and then by taking up a little of the mixture and putting it in a glass jar with about twice as much water. You can tell how far the paper has broken down. When it is well pulped you should not be able to see any recognizable particles of paper.

Using a Liquidizer

If you are using a liquidizer to pulp the paper, start by beating the wet newspaper for a few minutes, as I have already described; this will save time in the liquidizer. Break the damp paper into eight lumps, each about the size of a fist. Each lump will be the right quantity for a $1\frac{3}{4}$ pint (1 litre) capacity liquidizer. Put the first lump into the liquidizer and fill up with water. This mixture is only about 2 or 3 per cent fibres to 98 per cent water so it will not do any damage to the liquidizer.

Run the liquidizer for short bursts for about half a minute. If it is the sort of liquidizer which has a variable speed control you can run it at near top speed. You will be able to see from the absence of solid bits of paper that the job is done. Pour the pulp into a bucket and beat the remaining lumps of newsprint in the liquidizer.

A liquidizer runs at about 28,000 revolutions per minute which is why it makes such short work of pulping paper.

Pulping by hand.

Pulping with a liquidizer.

There are other ways of pulping which are slower but more suitable for larger quantities. Old washing machines, for instance, are quite suitable. If you have ever tried to get rid of one, you will know how easy they are to come by. One basic type has a stainless steel tank with an impeller fixed in the side. This impeller is driven by a belt from a motor outside the tank. Another type has a central agitator which reverses direction every so often. Experience will show how much paper you can put in for pulping but you may be able to run these machines at better than 4 per cent of solid matter.

Other Sources of Pulp

Almost any kind of paper can be pulped but avoid anything coated with plastic. Illustrated magazines are not very satisfactory as they, like newspapers, are heavily inked and made from low grade material.

At the other extreme is brown wrapping paper. This is very strong and usually has a low acidity. If you pulp brown paper on its own, naturally the paper you make will be the same colour but if it is mixed with white waste the result will be a pleasant light brown paper with good strength. Paper bags and carrier bags are also strong, and so are envelopes. If you include envelopes in your pulp, make sure they have not been sealed with sellotape; remove any transparent windows. Some commercial envelopes have a seal made from two coats of synthetic adhesive, one on the flap and the other on the body of the envelope, which stick when pressed together. This plastic adhesive will hang together in strings and it definitely does not make for good papermaking. The ordinary gum which you have to lick will not give any trouble.

Office waste paper such as old letters and file copies, as well as litho-printed price lists and similar documents, are good for pulping because the amount of ink on the surface is relatively small while the paper itself is of good quality. Paper used in dry-copiers is particularly good. The ink is deposited in the form of a light powder which breaks up easily and hardly discolours the paper you make. Dry-copier paper is also of good quality with fairly long fibres and low acidity.

Two other good materials found in offices are computer punch-tape and computer print-out paper. However, they have to be very strong to stand the high running speed of computer equipment and this unfortunately means that they are very difficult to break down. I have been successful with old bank statements, chicken-feed sacks and a lot of other materials.

Pulp for Pure White Paper

None of these sources of waste paper will give you entirely white paper, either because they are coloured or because they carry print. Personally, I don't mind that. Just as doctors say white bread is bad for you, and white sugar is bad for you, I maintain that white paper is bad for you. I am not suggesting it hurts your eyes or anything like that; it's just that if more waste paper is to be recycled commercially in future we will all have to get used to paper that is not pure white. If we can learn to accept a standard that is different rather than lower in future, many benefits will come from using more recycled paper. Less trees will be cut down, our import costs will be lower and less energy will be consumed than in the manufacture of paper from virgin pulp.

For your own rather more modest recycling operations you will find one source of pure white waste paper. The smaller printers often cannot get rid of the offcuts they trim from what they print. Sometimes they even have to pay extra to the local authorities to have it removed. If you know a jobbing printer he may be glad to let you have his offcuts.

Many of the materials I have suggested may need to pass through the liquidizer a second time to be completely broken down because they are much stronger than newsprint. A second beating, rather than a longer first beating, seems to be more successful in breaking down hard papers.

Commercial paper mills are supplied with their raw materials by pulp mills, usually in Scandinavia or North America. After the wood has been chemically or mechanically broken down, the pulp is dried and pressed into flat sheets for despatch to the paper mills. At this stage the fibres in the pulp have not been beaten. It is the job of the papermaker to beat the pulp in his mill because the length and degree of beating will determine the final nature of the paper. From the beating stage the pulp passes more or less directly to the paper machine. It is therefore not normally possible for an amateur papermaker to get pulp ready for immediate use. He may be able to get 'half-stuff' which he will have to beat for some time before the fibres are in the right condition for papermaking.

4
MAKING PAPER

When you have finished preparing the pulp you will have a bucket containing about 2 gallons (8 litres). If you made the pulp by hand, it will be more concentrated and you must add water to bring it up to 2 gallons (8 litres). In either case, give the mixture a good stir and pour just over half of it into the vat which you are going to use. The other half of the pulp can be left in the bucket; it will be needed for topping up the mixture in the vat from time to time. Each sheet of paper you form will remove some of the fibre content from the vat, making an occasional top-up necessary. Add $3\frac{1}{4}$ gallons (15 litres) to the pulp in the vat and give it a good stir. Fairly frequent stirring will be necessary as the fibres tend to settle.

Now take the mould, mesh side uppermost, and lay the deckle on top of it. Hold the mould and deckle together firmly with your hands on the two short sides, thumbs on top and fingers underneath. Hold them vertically over the vat and dip the lower edge into the pulp at the far side of the vat (Fig. 1). Let the mould and deckle down until the pulp completely covers them, then turn them to a level position just under the surface by bringing the lower edge towards you (Fig. 2). If you

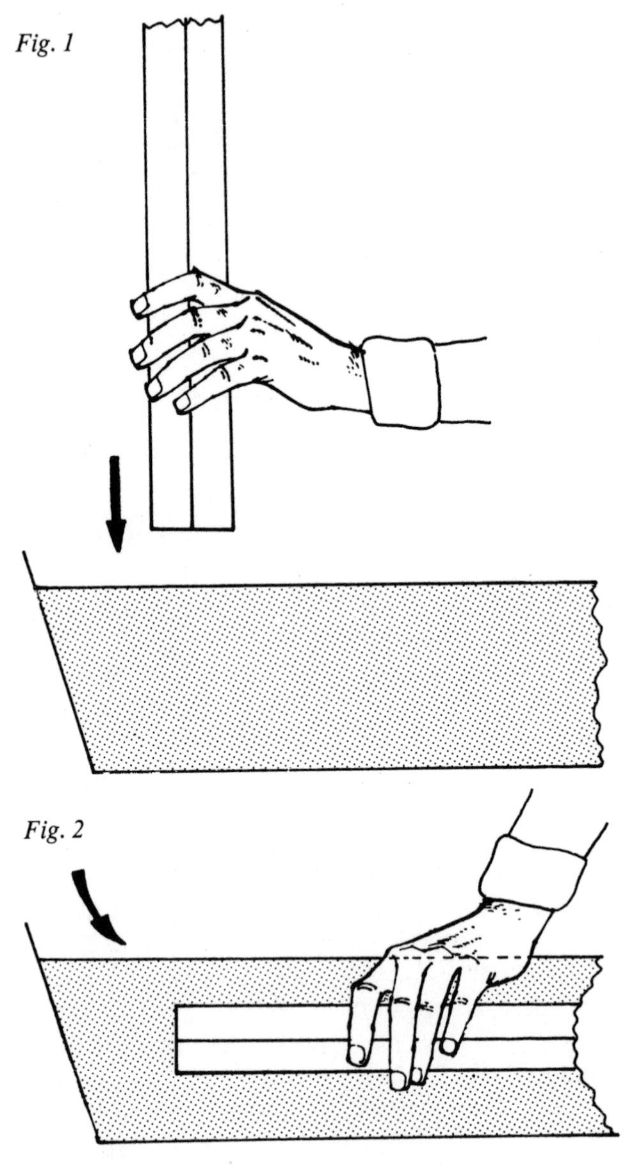

Moulding.

Fig. 3

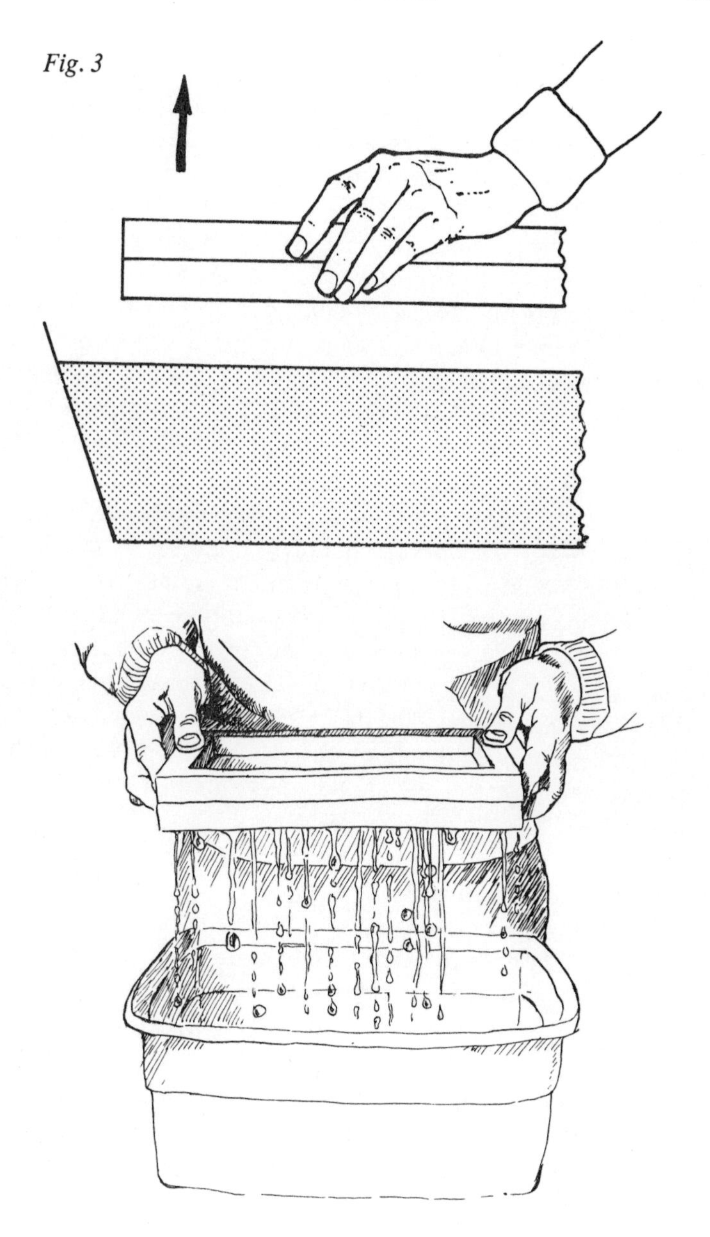

managed to follow that last bit of instruction, the mould and deckle will now be submerged with the deckle on top and the mesh covered face of the mould uppermost under the deckle.

Let the mould and deckle rest for a moment under the surface and then pull them straight upwards. Pull quite swiftly but steadily (Fig. 3). You will notice suction resisting your efforts to lift them. This suction helps to pull the fibres on to the mesh and form a well-made, even sheet of paper.

When the mould and deckle come out of the vat there will be quite a runny mixture of pulp on top which will become stiffer as water drains through the mesh. For a few moments you will be able to shake the mould from side to side and front to back. This helps to arrange the fibres but it is important to stop as soon as the pulp has lost enough water to become too stiff to move easily. Further shaking after this moment will only spoil the formation of the sheet.

Lay the mould and deckle down for a moment on a level surface; the corner of the vat will do, but take care that they do not fall into the vat. Now you can take off the deckle; it has done its job of holding the pulp on the mould while it was still very runny. Remove the deckle carefully to avoid drips of water falling on to the pulp on the mould. The layer of pulp covering the surface of the mould will be about $\frac{1}{8}$ inch (3mm) thick.

Couching the First Sheet

You will now be ready to couch the first sheet. If you have provided a drip tray to catch surplus water from the couched sheets, put that beside the vat. If not, use the draining board. Lay down two strips of wood on the tray or the draining board and put one of the boards of the press across them. Cover the board with two damp couching cloths.

Hold the mould over the press board with both hands at the short sides. Turn it over so that the pulp is underneath and press it on to the couching cloth, starting with the edge furthest away from you and pressing the rest of the mould gently on to the cloth until it is resting flat. The remarkable thing is that the pulp does not fall off when the mould is inverted; on the other hand, once the pulp is pressed on to the couching cloth it tends to stick to the cloth and free itself from the mould. You will have to help the first sheet by rubbing the mesh with your hand as the mould lies on the couching cloth. The mesh will become lighter in colour, showing that the layer of pulp is transferring to the cloth. You can pull the mould carefully up, starting with the near edge. Run a finger over any part of the mesh where the pulp is still sticking to it.

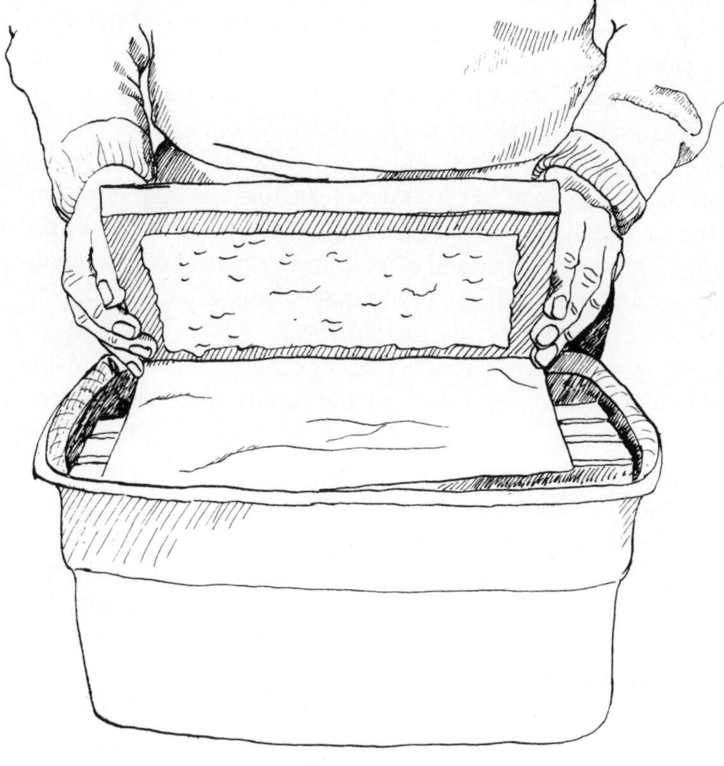

Couching.

This first sheet may look a bit of a mess. I cannot be sure why, but it usually happens that the first two sheets are always more difficult to couch than the remainder. Probably it is because as you build up a good mass of damp sheets of paper and cloths on the press board the later sheets have a good wet bed to lie on. Anyway, don't worry about the appearance of your first sheet – go straight on to make another in exactly the same way.

Making Subsequent Sheets

Cover the first sheet you have made and couched with another couching cloth. Start to lay the cloth down at one edge and stretch it slightly as you cover the rest of the sheet to ensure that there are no wrinkles in the cloth. Now you can mould another sheet from the vat in the same way as the first and couch it on to the press board. You can make about twenty sheets at a time to build up a 'post' on the board. After the second or third sheet you will see that the pulp is getting thinner because you have taken quite a bit of fibre out of the vat. Top up with a cupful of pulp from the bucket containing the second half of the pulp you made. You will need to top up after every sheet or every second one.

As you get more practice at couching you will find the action is rather like using an old-fashioned rocker blotter. You will put down the far edge of the mould, roll the mould on to the cloth until the centre and near side of the pulp are in contact with the cloth and then start lifting up the far side of the mould in one continous movement. You may even find it helpful to pack a wad of paper under the first couching cloth on the board to build up the centre just a little into a hump.

Thick or Thin Paper

The thickness of the paper will depend partly on the thickness of the pulp mixture in the vat and partly on the speed with which you handle the mould and deckle in the pulp. You will be able to make very thin sheets with practice but it is best to make your first paper thick because it is so much easier. If you can clearly see the pattern of the couching cloth through the sheet you have just couched, this is a sign that the paper is too thin. When you have no more pulp left for topping up and when the mixture left in the vat is too thin to make a good sheet, you can safely pour it down the drain or save it for the next batch of paper you make.

Recycling Unsuccessful Sheets

If you have any reason to be unhappy about any sheet you have made, you can return it to the vat. Turn the mould over, so that the pulp is underneath, and lay it on the surface of the pulp in the vat. The pulp will wash off the mesh quite easily and can be stirred into the contents of the vat. You can clean a bad sheet of pulp off a couching cloth in the same way. Don't be too ready to throw away what you have made: I often find that what looks like a bad sheet turns out to be quite respectable once it has been pressed and dried.

The quantities I have suggested will be enough to make between sixteen and twenty-four sheets, depending on how thick you make each sheet.

Pressing

When you have couched the last sheet on to the post, cover it with two couching cloths and put the other press board on top. Put two battens under the bottom press board and the other two across the top one. Push a bolt through the hole in each bottom batten and through the hole in the upper one and put on the washers and nuts. Tighten each nut in turn, a little at a time to get an even pressure all round. When you can tighten the nuts no more, leave the press for about five minutes and then open it up. It is not necessary to leave the paper in the press for long; it is the amount of pressure which matters, not the time spent under pressure.

If you have adopted my suggestion for using G clamps to close the press instead of nuts and bolts, the operation will of course be much the same as I have just described.

When you have opened the press, peel off the top two cloths to expose the first sheet. It has reached its final thickness and looks very much like a sheet of paper but it is still very damp and has to be handled with care. To avoid damage, keep it on its couching cloth; peel the cloth off the post, taking care to avoid picking up the edge of the sheet of paper underneath. If you start peeling at one edge of the cloth and find the sheet below sticking to the cloth, try another edge. Pressure with a finger on the cloth will help to free it from the lower sheet of paper.

Lay the couching cloth with its paper on top on to a sheet of newspaper and take off all the other sheets in turn. Laying them on newspapers will help to dry them more quickly. After an hour or so, turn each cloth over and lay it on a fresh sheet of newspaper so that the paper is next to the newspaper. It should now be possible to peel back the couching cloth from the paper you have made. All the sheets of paper can now be left to dry on newspapers. How long this will take depends on a variety of factors. In a warm room it may take about three hours. In a cold room it could take twelve hours. You can speed up the process by laying the sheets on a radiator or stove. You can even iron them with a warm, but not hot, electric iron. However, I always find the paper dries more evenly and with less risk of cockling if it is allowed to dry slowly. If any sheets do cockle anyway, you can try dampening them and leaving them under a weight for an hour or so.

When the sheets are virtually dry, lay them one on top of the other and put some books on top for a weight. Ten fairly thick books will be sufficient and they should be left for two or three days.

I mentioned before that it is best to clean the equipment

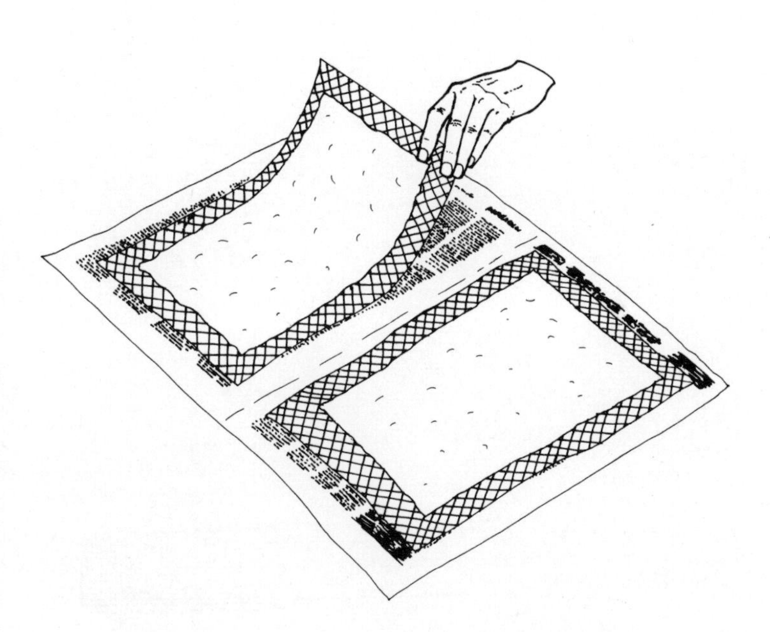

Laying couching cloth and paper on newspaper.

immediately after making a batch of paper, to prevent bits of old, dried pulp getting into the next making. It is amazing how impurities manage to get into paper. Dust and earth are also very tiresome; if a tiny speck gets into a sheet of paper when it is being made it will dry out with the paper and form a small lump in the surface. When it is rubbed or scratched it will break up and ruin the sheet.

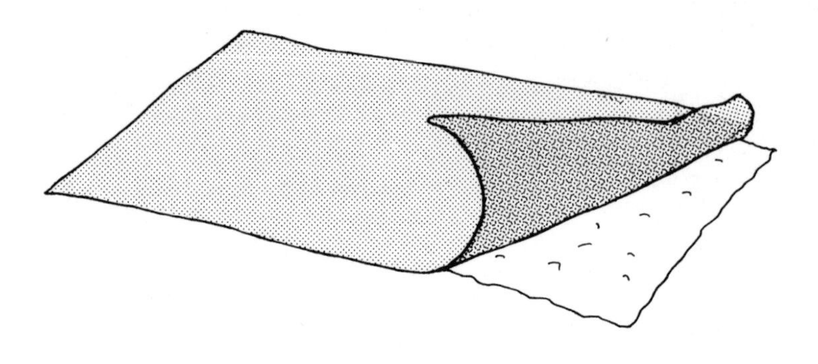

Peeling back couching cloth.

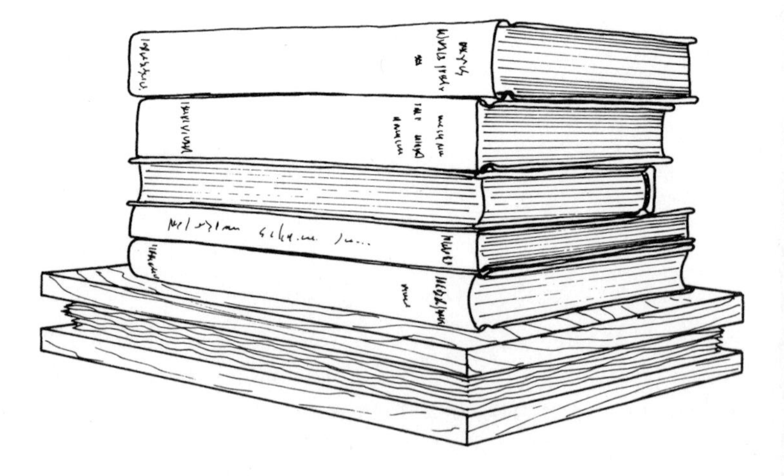

Weighting pile of damp paper.

Deckle or Trimmed Edges

The sheets you have made will have rough edges because some of the pulp will have flowed between the mould and the deckle during moulding and because of contractions in the fibres during drying. These rough edges are called 'deckle-edges'. If you don't like them you can guillotine the paper or cut off the edges with a sharp knife and a rule, but many people prefer to keep them. In fact, they used to be insisted on for high quality writing paper. Having a deckle edge on your stationery showed that you could afford the best. As machine-made paper became more general, machines were developed to give each sheet a bogus deckle edge, so creating the illusion that the paper was hand-made.

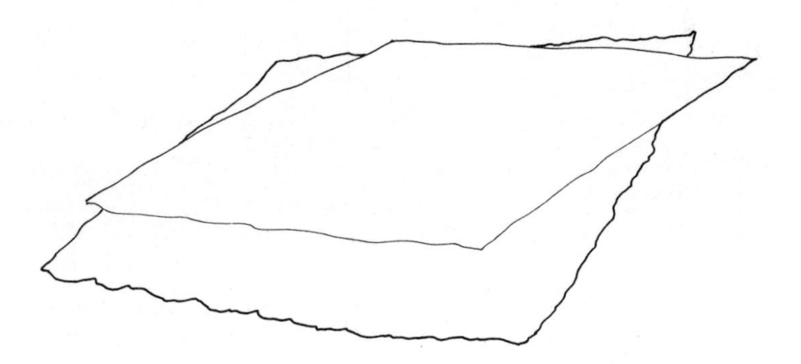

Deckle edges.

The drying process will have left the paper with a fairly rough surface. If you want to get rid of this you can smooth the paper with a warm iron or a rolling pin. In the next chapter I will describe a method for getting a glass-smooth surface.

Single Sheet Forming Method

You may want to make just one sheet of paper. This will probably be because you want to try out a new mix of pulp without wasting too much of it or because you have only a very small amount of pulp. The single sheet forming method will allow you to do this. It is a process often used in laboratories to test samples.

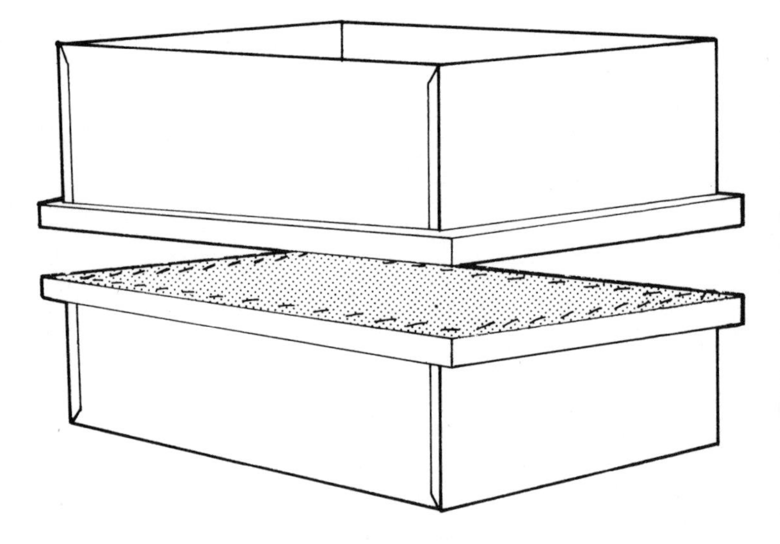

Single sheet former.

A mould and deckle are needed, similar to the usual set but both about three times as deep. You can easily adapt your normal mould and deckle. Take a strip of sheet tin or aluminium about $2\frac{1}{3}$ inches (6cm) wide and fold it so that it fits inside the deckle on all four sides. Make the same for the mould, fitting this lining below the mesh.

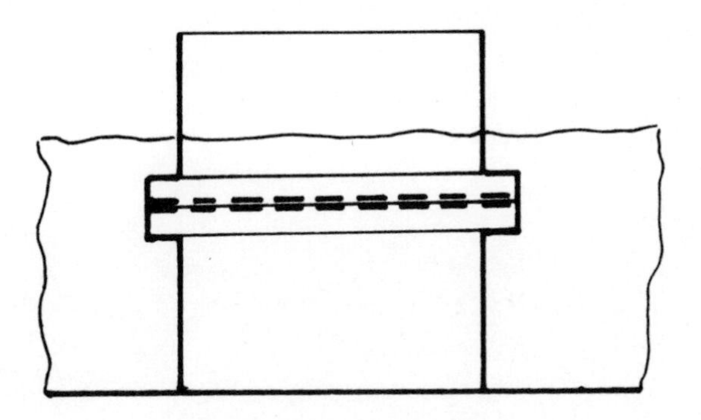

Single sheet former *in situ*.

You need a basin or sink filled with water to a depth of 4 inches (10cm). Have the pulp you are going to use ready, put the deckle on top of the mould and lower the two into the water so that the water rises over the mesh and the adapted mould rests on the bottom. The extended deckle will now hold a reservoir of water. Pour the pulp into this reservoir and lift the mould and deckle out of the basin, shaking them slightly to make sure the pulp and water are well mixed. As the mould and deckle leave the water, the pulp will be pulled on to the mesh by suction and excess water will drain through the mesh.

You will now have a layer of pulp on the mesh which can be couched immediately in the usual way or left on the mesh to dry for a while. It can be left there until entirely dry and then peeled off. If you are not satisfied with the result you can scrape the pulp off the mould and try again. This method is slower than the normal process and does not give such well-formed sheets but it is useful for making one-off sheets of special materials.

5
SPECIAL EFFECTS

Having tackled the basic process of papermaking, there are lots of interesting ways of creating special effects in the paper you make. There are certain inexpensive and readily obtainable materials referred to several times later in this chapter, so I am going to list them now with a few comments.

PVA adhesive. This is a synthetic resin adhesive, soluble in water when wet, but impervious to water when it has dried. It can be bought from art shops and is a white liquid with a consistency like thick cream. It can be thinned with water and spreads easily with a brush or palette knife. It is important to wash brushes out immediately after use, before the adhesive has time to set. Sometimes, bottles of PVA adhesive are hard to open because the adhesive has set around the cap. Running hot water over the top helps to free the cap.

Clear Gum. This adhesive is soluble in water both when wet and when dry. It often comes in a bottle with a rubber top incorporating a spreader; the mouth of this spreader tends to get bunged up with dried gum, so before using it, wash the spreader under a hot tap to clear it. Clear gum comes from art shops and stationers.

Gelatine. Gelatine is widely available in powder form from grocery stores.

Spreaders. The adhesives will need to be spread evenly. A painter's palette knife with a flexible blade is a useful tool but plastic spreaders sold in stationery shops (and sometimes given away with glues) are equally good.

Perspex. Perspex is one of the brand names for acrylic sheet; another version is called Oroglas. A sheet about $\frac{1}{8}$ inch (3mm) thick will help you to get a glass-like finish on paper. Unfortunately, it is rather expensive but a piece about twice the sheet size of your paper will give you a chance to see if you like the effect it gives. Formica laminate has rather the same result; you may have plenty of this covering your kitchen furniture.

Polyethelene sheeting. Polyethelene (better known as polythene) sheeting of 500 gauge has a number of uses and it is quite easy to get. It is widely used for wrapping, and gardeners cover plants with it to protect them.

Colouring the Paper

The simplest of all ways of decorating your paper is to colour it. Various substances can be added at the pulping stage; don't add colour to the vat because it will not mix evenly with the pulp. The easiest form of colouring agent to use is powder paint. It is simple to add to the pulp and it does not stain the equipment or your clothing. Start with 1-1$\frac{1}{2}$ teaspoonsful of colour to each pulp load in the liquidizer.

The addition of some paper which is already strongly tinted will colour the mixture. You can also use cooking dyes which are soluble in water (for instance, cochineal) but these colourings become rather expensive on the scale of paper making. John Mason, in *Papermaking as an Artistic Craft*, warned against using dyes because they would discolour couching felts and other equipment. I can't say I have found this too much of a problem with the sort of synthetic couching cloths I have recommended and I have used fabric dyes quite successfully to colour pulp.

Duplex sheets with leaf.

Duplex Sheets

Some of the most exciting and attractive effects come from trapping leaves and grasses in the paper. I have even heard of dragonfly wings being used for decoration, though this seems to require the catching and killing of rather a lot of dragonflies, which is a pity. To make a duplex sheet, first form a very thin sheet, couch it in the normal way and then lay the leaves or grasses on top of it. Next mould another thin sheet and couch it straight on top of the first sheet. When the double sheet has been pressed and dried, the outline of the leaves and grasses will show through the paper. Similar effects can be obtained with pieces from paper doilies. Examined at close quarters, a doily will be seen to be a pattern of small flower shapes which can be cut out very easily and rearranged as you want them. The doily should be stained with colour first so that the pattern shows up well in the paper.

You can repeat Congreve's security watermark experiments by duplex couching sheets of different colours if you have a separate vat containing pulp of the second colour.

Thread Pictures

I don't care for this idea myself but you may like to give it a try. Make the first sheet of a duplex, lay down a thread in the form of a pattern and then couch the second sheet on top of the thread.

Additions to the Pulp

Grasses and seeds can be added at the pulping stage and they will appear in each sheet of paper you make. It won't, of course, be easy to write on the paper because it will have lumps in it, but interesting patterns can be created. Mixing torn-up coloured paper tissues with the pulp in the liquidizer makes fascinating effects. The makers of the tissues have tried to give them 'high wet-strength' by adding various chemicals. It is obviously important that the tissues should not break up immediately they get wet. As a result, the tissues in the liquidizer will not break down as completely as the pulp. When you make a sheet of paper it will be coloured by particles of the tissues.

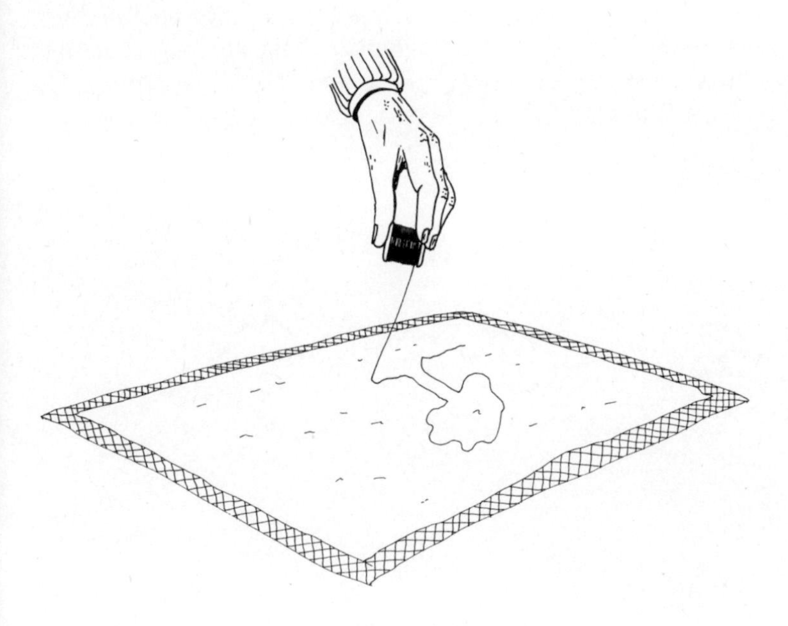

A thread picture.

My own favourite effect comes from cutting coloured wool into very short lengths, not more than 1/10 inch (2mm) long and adding these bits to the pulp in the liquidizer. The wool will break down into thin hairs and will be scattered throughout the sheets of paper. This gives an effect which papermakers call 'silurian'. You can get similar results from using scraps of cotton cloth.

Sizing

Sizing is an operation to seal the surface of the paper so that water-based paints and inks will not run. Anyone who has tried to do a newspaper crossword with a fountain pen will know what happens when you write on unsized paper. Ancient Chinese artists, though, preferred their paper to be left unsized. Using pens which were the ancestors of the modern felt-tips, they took advantage of the absence of size to vary the thickness and weight of each line according to the speed with which the pen travelled over the paper.

Sizing technique.

Engine-sizing

There are two forms of sizing, known as 'engine-sizing' and 'tub-sizing'. Engine-sizing is carried out in the vat while the paper is moulded. The size added to the pulp used to be rosin which was followed by aluminium sulphate, and the chemical reaction between these two created the sizing effect. You can engine-size your own paper by adding a preparation called *Aquapel 5*, made by the Hercules Powder Company. It is stocked by one or two art and craft shops, and can also be obtained from Paperkit Ltd (for their address, see page 91).

Aquapel is added to the mix after it has been pulped and just before it is poured into the vat. Experiment will be needed to determine the right amount for your taste but as a general rule 3 teaspoonsful of *Aquapel* are needed for over 2 oz. (50g) dry weight of material. The makers of *Aquapel* rightly err on the side of caution and give it a very short shelf life but I have found no problem in using *Aquapel 5* which is two years old. It is important to keep it in a cool place.

Tub-sizing

For my part, I prefer to tub-size. This means a separate operation because the paper has to be passed through a bath of size after it has dried and must then be allowed to dry again. My reasons for favouring tub-sizing despite this disadvantage are as follows: the gelatine used in tub-sizing adds strength to the paper as well as sealing it, whereas the *Aquapel* gives no extra strength; *Aquapel* takes fountain pen ink less well than gelatine; a heavy application of gelatine can be used to give a glazed effect if that is wanted. Finally, I prefer tub-sizing because when I am making the paper I cannot be sure what I will eventually be using it for and whether I will want it sized. Printing on hand-made paper, for instance, looks better if the paper can be dampened and this is easier when it is unsized.

The simplest form of size tub is a shallow dish a little larger than the sheets of paper to be sized and about 1 inch (2·5cm) deep. The size consists of 1-2 teaspoonsful of gelatine powder dissolved in 1¼ pints (¾ litre) of warm water. Pour this mixture into the shallow dish. Each sheet has to be dipped into the size and then laid out to dry.

The dipping has to be done with care because the paper tends to break up when it gets damp. You can keep one short edge of the paper out of the size and use this edge to hold the paper when you pull it out. A safer method of handling the sheet while you are sizing it is to lay it on a couching cloth and drag the cloth through the size. Hold the cloth with the paper on it over the size for a few moments after you have taken it out to allow surplus size to drain back into the dish.

Lay the paper on to a sheet of stiff card or a sheet of polythene and press the back of the couching cloth to push all bubbles from the centre to the edges. Then peel off the couching cloth and leave the paper to dry.

A glazed surface can be given to the paper by using a stronger size mixture or by giving the sheet a second dipping in size when it is dry again. Another method of glazing is to brush on two or more coats of PVA adhesive. This is easier, as well as more economical, if the PVA is diluted with an equal measure of water. You must allow each coat to dry before the next is applied. When you put on the last coat, leave it for a few moments and then lay a piece of polythene over it. This will improve the glaze if you leave the polythene on until the PVA is dry.

Starch

My friend Bruce Glasser adds laundry starch to his pulp. This makes the paper much stiffer and gives it what papermakers call a nice rattle. Trial and error will tell you how much to use to suit your purposes but 1 or 2 tablespoonsful should be about right. I have recently started to use it for making writing paper and I find it gives a very nice surface. The type

I have tried has to be mixed first with a small amount of cold water and then with a larger volume of boiling water. I then pour this mixture straight into my pulp.

China Clay

China clay, or kaolin, is a mineral found mostly in Cornwall. It is mined in open pits and washed and screened before being sent to the paper mills. Some of it is pumped in slurry form in large diameter pipes from Bodmin Moor down to the river Fowey and loaded straight on to ships. It is used either as a coating for paper, when it produces the shiny effect found in glossy magazines, or as a filler.

I have not so far succeeded in my efforts to reproduce the coating effect on hand-made paper. However, gelatine size and PVA can give quite a good shiny surface, so coating with China clay is not so necessary.

As a filler, China clay fills the microscopic gaps in the paper between individual fibres. It makes the paper smoother, denser, whiter and more opaque, and for all these reasons it is then more suitable for modern printing methods. China clay also saves pulp, since it can be added in a ratio of up to 10 per cent of the dry weight of the pulp.

If you can get some China clay you can try it for yourself by stirring it in with a little pulp and then adding the mixture to the main body of the pulp. It must be well stirred to get rid of all lumps, and remember not to add more than 10 per cent of the dry weight of pulp. A greater proportion will make a very weak sheet.

Surface

Some people like the surface of paper to be fairly rough; for drawing in pencil, for example, a rough surface is important because the indentations catch the lead of the pencil and give the drawing its character. Artists call it 'tooth'. For some water-colourists, a rough surface is also important.

Personally, I like the rougher surface for printing on because it gives a really bold appearance to the type on the paper.

If you leave the paper you have just made drying on a sheet of newspaper or a piece of cloth it will have a rough surface when it dries because of the uneven contractions that take place as the drying is taking place. If you prefer a smooth surface, however, it is easy to achieve. The newly-made paper must be taken from the post after pressing and laid out, sheet by sheet, on a smooth surface. It must be left on the surface until it is quite dry, when it can be peeled off. It will then have a surface identical to the surface it was laid on. The other side, uppermost during the drying, will be less smooth but unless the pulp contained a lot of rough particles it will be quite even.

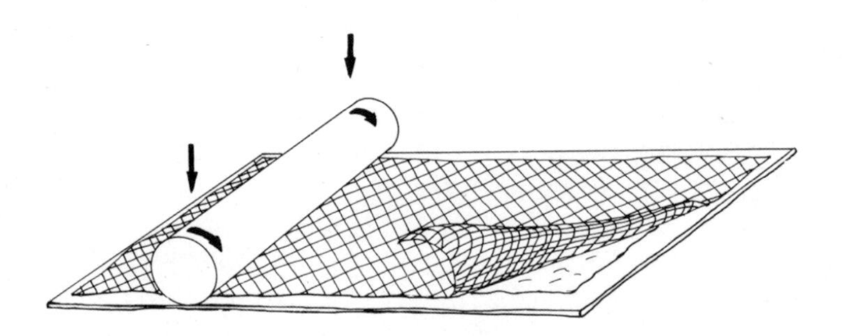

Surfacing on cardboard.

The ideal drying surface for a very smooth finish is a sheet of perspex or formica. You may have formica work-surfaces in your kitchen. Even the sides of kitchen units will be quite suitable because they usually have a smooth exterior face. For a good, though not quite so smooth, surface you can use sheets of cardboard; unless these are very thick they will tend to cockle with the moisture from the drying paper so after use they must be dampened and stacked under a weight to make sure they recover their shape. By doing this you can use the cardboard sheets many times over.

I was disappointed to find that window glass was not suitable for this process of smooth-drying. It seemed that the easiest answer to the space problem would be to plaster the paper on to the windows and let it dry there, but I found it was very difficult to peel the paper off without tearing it. Space, of course, is a problem; the advantage of using cardboard sheets is that you can punch a hole in one corner of the cardboard and string a line through several sheets, or you can clip the sheets to a washing line. If you only have space to dry two or three pieces of paper at a time you can leave the post of damp sheets covered with a wet cloth until you have worked your way through all the paper.

Heat can be used to speed up the process of drying, but it must be even; a fierce direct heat which might be stronger on one part of the paper than on another will cause cockling. A warm room is preferable to an electric fire giving localized heat.

When the paper is laid on to the drying surface, any bubbles of air which get trapped between the paper and the surface will lead to rough spots in the paper. When you lay the paper on to the drying surface, leave the couching cloth on top of it until you have pushed all bubbles out to the edges from the centre by hand pressure. You can also use a rolling pin, working outwards from the middle. You will now be able to peel off the couching cloth. Rub a finger round the edges to make sure the paper stays on the surface and does not follow the cloth away.

As I mentioned earlier, it is important to leave the paper on the drying surface until it is quite dry. This might take six to eight hours in a warm room. If you peel it off too soon it will cockle. There is nothing new about this method of drying; the Japanese have laid their paper out in the sun on smooth wooden boards for centuries, and elsewhere the wet paper has been pressed on to clay walls to dry.

If you have access to an etching press you can give the paper a smooth surface by passing it through the rollers on a flat metal or plastic sheet when it is dry. Bruce Glasser gives his paper a patterned surface by putting it through his etching press backed by a piece of cloth with a strong texture.

Printing on Hand-made Paper

I have a very small printing press, made by *Adana*, and a quantity of type. I can print an area of up to 4 x 2 inches (10 x 5cm) and this allows me to make my own letterheading. I can also print small cards with short messages. Because hand-made paper is less heavily pressed than commercial paper, and therefore softer, it is much easier to get a good impression than it is on normal papers. The effect is particularly good if one prints on paper which has been slightly dampened. I usually get the right amount of dampness by wetting a few couching cloths and squeezing most of the moisture out of them. I then build up a stack with a couple of sheets of paper between each damp cloth. I leave the stack standing for a few hours under a weight to make sure the dampness is evenly distributed.

You can take a print from a lino-cut by dampening the paper in the same way. I enjoy making relief prints by arranging flat objects such as metal washers in a pattern and inking the surface with a roller. The paper can then be laid on to the inky pattern and rolled hard to pick up the design.

The softness of hand-made paper also means that it is possible to make quite an effect by blind blocking – that is making an impression on the paper without inking the image from which the impression comes. After putting the damp paper over the relief which is going to provide the impression, put several layers of newspaper or cloth on top of the paper and then press hard. If you have a copying press of the type I described earlier, so much the better. If not, you can use the press made for expelling water from your paper, or even make a sandwich of two boards held together with a G clamp.

The effect of this process of taking a relief impression is a picture incised into the paper; on the reverse of the paper there will, of course, be a raised pattern. You can go even further and combine the two, with an incised and a raised design side by side, or an embossed blind pattern coupled with normal printing.

The most delightful pictures can be made by rolling ink over natural objects, such as leaves, grasses and feathers, and then taking an impression on dampened paper. These techniques and many others are often used in schools but they are all much more effective on hand-made paper than on commercial materials. This is because so much is added to the print by the texture and softness of the hand-made paper, together with the added advantage of being able to print on unsized paper, which can give so much more force to the printing ink. Anyone involved in screen printing will find that his prints have much more impact because of the way the ink is sucked into the paper to give really bold colour effects.

Envelopes

If you open an envelope out to its full extent you will see that it is usually in the shape of a four-pointed star or a diamond. During the manufacturing process, three of the points of the star are folded over and glued together; the fourth has a gummed edge put on it and is then left open to act as the flap. You can of course cut your paper in the same shape to make an envelope. You can also mould paper in the shape of an envelope by cutting a template out of stiff card and laying it inside the deckle when you are making the paper. The pulp will arrange itself on the mould in the envelope shape. This technique can also be used to make paper into other shapes. The card used to make an outline should be sealed with two coats of PVA adhesive, otherwise it will disintegrate after a few immersions.

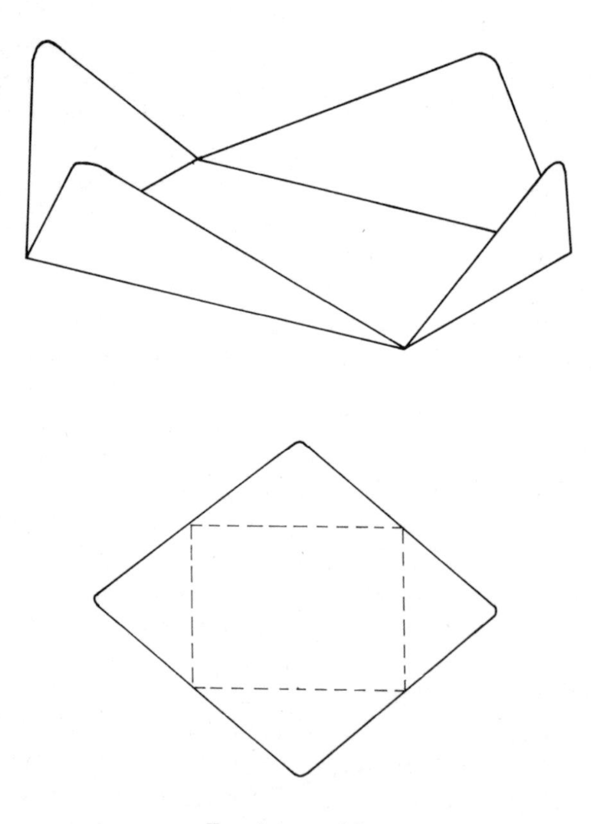

Envelope making.

Clear gum can be used to stick the three sides which are to be folded over and also to pre-glue the fourth side. Unless the paper from which the envelope is made has been well sized it will be necessary to lay down a coating of PVA for the gum. If this is not done, the gum will soak into the paper and not make a good stick. When you want to use the envelope, the fourth side can be moistened to activate the glue.

Economy Labels

Most of us get more mail nowadays than we send out, if only because the postman brings in so much advertising material. No one is short of envelopes, though many of them will make a good source of raw material to pulp, as I mentioned earlier. If you have any left over, you may prefer to use them again with economy labels. You can make your own economy labels by coating paper with a thin film of clear gum. As in envelope making, the paper must be well sized or coated with PVA before spreading the clear gum.

Laminating

Commercial papermakers produce thick card by arranging several cylinder machines (of the type invented by John Dickinson in 1809) in a line. The web of newly-made paper from the first machine passes over the second and is pressed on to the second web. More layers are added before the thick multiple web passes through the pressing and drying section of the plant.

You can make your sheets thicker by duplex couching in the way I described earlier, but this process does not normally work for more than three layers. Anything thicker gets spoilt in the pressing stage. It also takes a long time to dry a very thick sheet. Dried sheets of paper can be laminated into any thickness you want by coating the sheets with PVA and holding them together under pressure or under a heavy weight until the adhesive has dried.

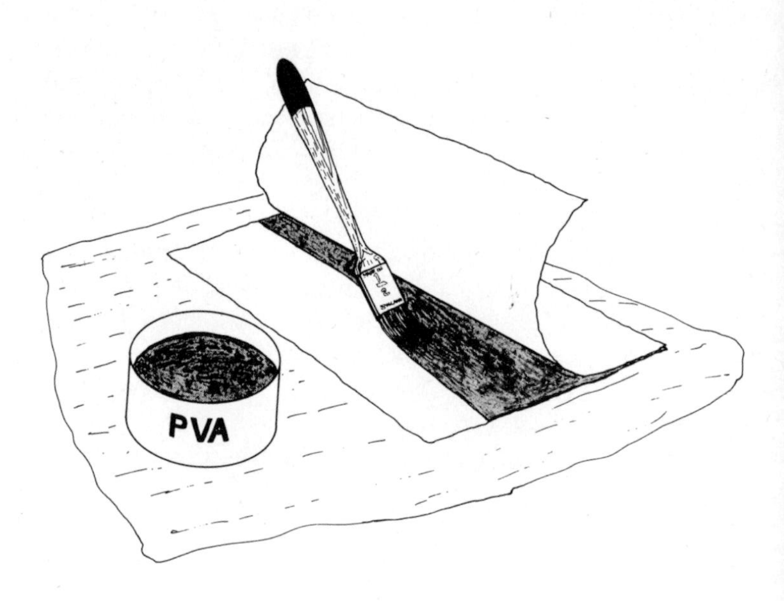

Laminating.

Watermarks

You can easily make a watermark for your paper. You must first decide on a design. This ought to be simple as it would otherwise be difficult to read and would spoil the surface of the paper. It is probably easiest to start with your own initials. Draw out the design on a sheet of paper and then lay the mould face down on the paper and trace the design on the mesh with indian ink or a fine felt-tipped pen. When you look at the upper face of the mesh the design will be a mirror image with the characters in reverse. This is necessary because the watermark will then be the right way round when you have made paper and dried it on a smooth surface.

Having traced the design on the mesh you can weave a thick thread of cotton, or a doubled strand, in and out of the holes along the lines of the design. An alternative method is to paint in the design with an adhesive which is water-resistant once dry. PVA will do.

Marbling

Marbling is one of the most exciting and attractive ways of decorating paper. Spots of oil-based colour are floated on a water surface with which they will not mix, nor will they mix with each other. A pattern can be created by dragging the colours across the surface and the pattern can then be picked up by laying a sheet of paper on the surface. When the paper is peeled off, it will carry the marbled impression. A fresh pattern must be laid down for the next sheet of paper.

Marbling has been used for centuries to decorate the covers and end-papers of books. Marbled paper can be used to cover lampshades, note-pads, wastepaper baskets, trinket boxes, and wall panels. The most controlled pattern can be achieved by using Carragheen Moss (an Irish seaweed) mixed with water, as the ground or surface which takes the colours. With Carragheen Moss and special colours thinned with ox-gall, professional paper marblers can repeat the same intricate pattern over and over again, with every sheet almost identical to the last.

Amateurs will find a gelatine based ground and ordinary oil colours easier to use, though they will have less control over the development of the pattern.

You will need a shallow metal or plastic tray, about 14 x 10 inches (35 x 25cm) in area with a depth of 1 inch (2·5cm). Put one dessertspoonful of gelatine in this tray and pour over it $\frac{1}{2}$ pint (250ml) of very hot but not boiling water. Stir well until all the gelatine has dissolved. Next add $\frac{1}{2}$ pint (250ml) cold water and let the mixture cool to room temperature. When it is cool, the ground should still be liquid, not set to a jelly. If it has turned to jelly, add a little warm water to thin it.

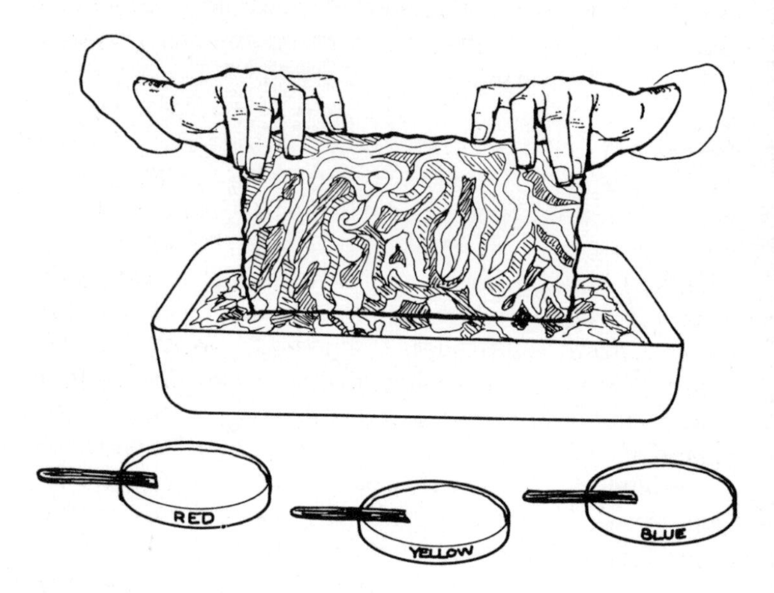

Marbling process.

You will find it best to start with no more than three colours. Each colour will need its own small mixing dish; the top of a jam jar is perfect for this purpose. Squeeze out about 1 inch (2·5cm) of each colour and add white spirit until the colour has become roughly the consistency of single cream. You can test the colour by dropping a little on to the gelatine ground. If it sinks, the colour is too thick and more white spirit must be added. If the colour spot spreads to more than about 1 inch (2·5cm) it is too thin and more neat colour should be added to thicken it.

Any fairly thick paper will be suitable for marbling. White cartridge paper is ideal and so is brown wrapping paper, though brown paper, when it has been marbled, will look more subdued than white. It is the dazzling effects you get with white marbled paper that I particularly like. If you are going to marble your own home-made paper, make sure it is well sized first. If it isn't, not only will it tend to fall to bits in the marbling tray but it will also soak up the colours and make the contrasting edges of each colour spot less sharp.

Drop a few spots of each thinned colour on to the surface of the ground and form a pattern by dragging the colours across the surface with the point of a knitting needle, a long nail or a sharp stick. When you have a pattern you like, slowly lay a sheet of paper on to the surface and leave it for a moment. Peel it off by one corner. Hold it to drain for a few moments and run tap water over it to wash off surplus gelatine. Then lay it, face upwards, on a sheet of old newspaper or a piece of perspex to dry.

Skim the surface of the ground with a piece of newspaper after taking each print and then you can repeat the marbling process.

Perfect Binding

Perfect binding is the technical name for a form of unsewn binding. You will see it used on many thick magazines and on almost all paperback books. The pages of a volume are laid one on top of the other in a stack and lined up to present a straight edge at the spine. Adhesive is then worked into this edge so that it makes contact with each sheet. When the adhesive is dry the book can be opened and every page will remain stuck in the book. That, at any rate, is what is supposed to happen. I expect everyone has had experience of a paperback which has been so badly bound that all the pages come adrift.

You may aspire to bind books in the traditional way, with large sheets folded two or more times to make sections which are sewn together and then encased in a cloth or leather-covered board. It is my ambition, though I doubt if I will ever

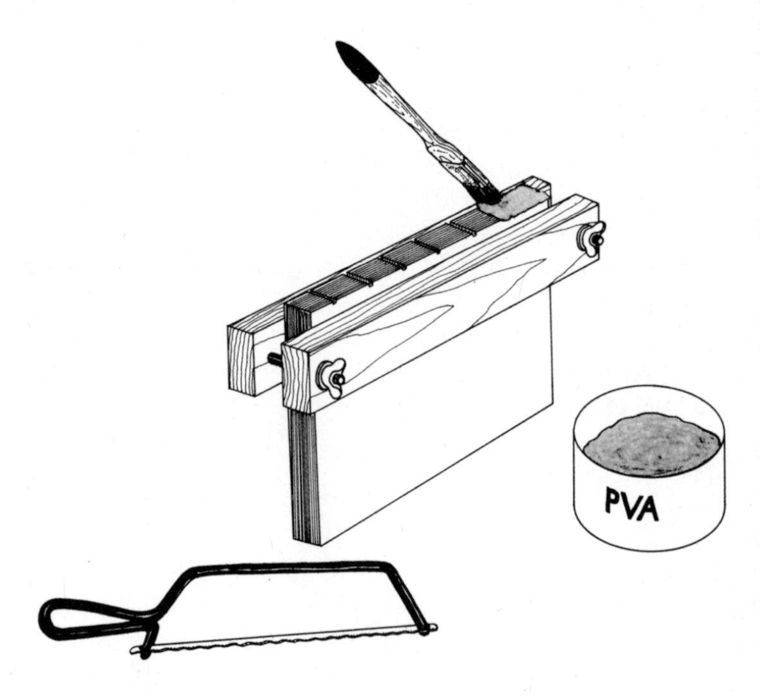

Perfect binding.

realize it, to write a book, print it on my own paper, and finally bind it myself. Until I can find the time and acquire the skill to achieve this, perfect binding makes a useful substitute. It is easy to do and needs nothing more than some PVA adhesive, a small saw or sharp knife, a length of fine thread and a gauze bandage.

First take the sheets you are going to bind and put them together in the form of a book. For covers, you can use thicker sheets or pieces of cardboard. Put one of these at each end of the book. Next take the book and knock it hard on a table top to get a good even edge and lay it down carefully to avoid disturbing the pages. Take two pieces of wood just a little longer than the spine of the book and lay one along the front cover and the other along the back. Clamp these two lengths of wood together at each end of the book or grip them in a vice. The edge of each piece of wood should lie about $\frac{1}{8}$ (3mm) in from the edge of the spine.

Cut about eight nicks with the saw across the spine to a depth of about 1mm. Use a spreader to rub PVA adhesive along the spine, pushing it well into the pages. Lay short lengths of thread in the nicks. These will help to ensure that no pages drop out of the book when it is finished.

When the PVA adhesive is almost set, adjust the two pieces of wood so that they are only about 1 mm from the spine. If the book is more than a few pages thick a strip of bandage will give extra strength. You must work the PVA well into this bandage.

There are refinements you can make to improve the appearance of a perfect bound book. The spine can be concealed by taking a strip of cloth or paper and glueing it to the spine. This strip must be the length of the spine and wider than it by $1\frac{1}{2}$ inches (4cm). The extra width allows for carrying the strip around the spine and on to the back of each cover board. You can also bind the sheets of the book without cover boards and glue on a wide bandage afterwards. The cover boards can be glued to the bandage afterwards and the spine concealed with a strip of cloth or paper in the way I have just described.

6

NATURAL MATERIALS FOR PAPERMAKING

Waste paper offers a wide range of material for making your own paper. The waste paper itself will of course once have been growing trees in a forest, natural vegetable matter. You may want to go one stage further back in the process of papermaking by using natural materials direct from the growing plant. Whatever the plant, you will have to carry out a process called digestion: the breaking down of the plant into its basic fibres and the removal of non-fibrous parts of the plant. In strictly non-scientific terms, you can think of a plant as being like a human being. It has bones held together with soft flesh. It is the flesh of the plant that you have to get rid of by digesting. The bones are the fibres you have to keep to make paper.

Cotton and linen rags have always been a source of raw material for papermaking; the same digestion process is needed to prepare them. It is important, if you are using rags, to choose only those which have not got synthetic fibres added to the natural material. Plants which are suitable include nettle stems and roots, cow parsley stems, rushes, reeds and coarse grasses, montbretia, gladioli, irises and potato stems. A couple of years ago I made some paper from lawn grass; you can still trace the summer smell of mown grass in the paper now. The plants should preferably be gathered in late summer or early autumn. The digesting process will reduce the bulk considerably, so collect at least two sacksful.

If you are going to try pulping rags, it is best to sort the rags into different colours and treat each colour separately. All buttons and other attachments must be cut off first.

Digesting the Raw Material

Whatever the raw material, the first stage of processing is digestion. The method most commonly used is cooking in caustic soda, though I prefer a cold process involving lime. If you use caustic soda, great care must be taken. You should wear rubber gloves to protect your hands and be particularly careful not to get any near the eyes. The material to be digested is put in a pot and covered with water. Caustic soda is added at the rate of two dessertspoonsful to 1¾ pint (1 litre) of water. The material has to be simmered for two or three hours to break down the plant and loosen the non-fibrous parts.

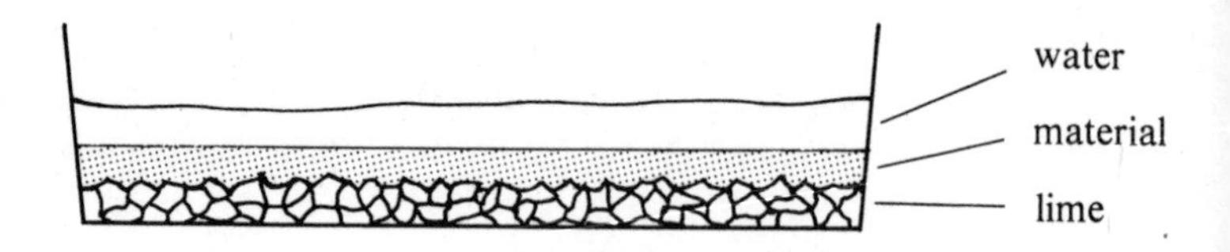

Digesting with lime.

Considering all the dangers of using caustic soda, not to mention the damage it can do to the surface of your cooker, you may prefer to try a safer, if longer, digesting process. This method needs builders' lime; I have used it successfully to break down cotton sheeting and banana tree stems. It should work equally well with other materials. A large shallow tray is needed. The bottom must be covered with about 1 inch (2·5cm) of lime; the material to be digested is laid on top of the lime and water poured in to cover it. Digestion takes about seven days. The mixture must be stirred each day but no other attention is needed.

Washing the Pulp

After seven days the material can be taken out of the lime tray and washed. The lime in the tray can be used again. The washing must be very thorough to remove all traces of lime. After this the material can be cut into small pieces, no larger than ½ inch (1cm) square, and pulped in a liquidizer or beaten by hand. Pulping or beating will take much longer than for waste paper. From time to time during the beating stage the stuff must be washed to get rid of the non-fibres. Make up a bag of mesh or coarse muslin to contain the pulp and hold the bag under a running tap while you squeeze and turn it for several minutes. The fibres will remain in the bag while the bits you don't want will be washed out.

An alternative method of washing the pulp is to lay it on the papermaking mould, using the mould with the under surface uppermost so that its sides form a sieve. Hold the mould under a running tap and turn the pulp with your hand under the water, taking great care to lose none of the pulp proper over the sides of the mould.

The length of treatment for each plant will vary so you must be prepared for a certain amount of trial and error. You can make a simple test to see if you have pulped and washed the material enough by making a piece of paper by the single sheet forming method. That way, you will not waste any pulp and you can form a sheet very quickly. You can iron it to get rapid drying. After inspecting the sheet you will be able to re-pulp it.

Cotton rag does not need much washing but it does need a great deal of pulping. Even then, it may be hard to make an even sheet because the cotton fibres tend to stick together in clumps in the water and it is difficult to keep them well dispersed in the vat. On the other hand, cotton is a very strong material and relatively easy to pulp. It can be mixed with other materials to give extra strength to the paper and this will lessen the problems of using cotton on its own.

The single sheet test will show you whether the pulp is ready for papermaking. Lumps or specks of darker material in the trial sheet will tell you that the pulp needs more washing to get rid of non-fibres. You may also consider that the sheet is too dark in its natural colour. In that case you will have to bleach the pulp before making it into paper. The pulp can be put into a glass jar with most of the water removed. Add a dessertspoonful of household bleach to $\frac{3}{4}$ pint ($\frac{1}{2}$ litre) of pulp and leave it to stand for an hour or two, stirring occasionally. As bleaches vary in strength, trial and error will once more be needed to establish exactly how much bleach to add to the pulp and how long to allow the pulp to stand after adding the bleach. All the bleach must be washed out before you make paper.

Pulp left for more than a day or two will start to smell. This is quite natural as it is vegetable matter, and it should not affect the quality of the pulp. A little bleach can be added to the pulp to stop it going off. You can also use the sterilizer sold for cleaning babies' milk bottles.

FURTHER READING

Unfortunately, many of these books are out of print but they are often stocked by Central Libraries. In London, most are in either the Library of the Victoria and Albert Museum or the Science Museum Library.

150 Years of Papermaking, J. Barcham Green, Barcham Green, Maidstone, 1960.

CIBA Review 72 – Paper, CIBA Basle, 1949.

History and Process of Paper Making, Paper & Paper Products Industry Training Board, Potters Bar.

Marbling, R.C. Akers, Dryad Press, 1976.

Modern Paper Making, R.H. Clapperton, Blackwell, 1952.

Papermaking, Dard Hunter, Knopf, New York, 1943.

Papermaking as an Artistic Craft, J.H. Mason Faber, London, 1959. Twelve By Eight Press, Leicester 1963.

Paper Making by Hand in 1967, J. Barcham Green, Barcham Green, Maidstone, 1971.

Paper Making in the British Isles, A.H. Shorter, David and Charles, 1971.

Papermaking Through Eighteen Centuries, Dard Hunter

Rudge, New York, 1930.
Printing for Pleasure, John Ryder, Bodley Head, 1976.

Some of these books, and materials and equipment for papermaking, can be obtained from:

Paperkit Ltd
Melbourn Bury
Royston, Hertfordshire, England

INDEX